Lecture Notes in Mathematics

Edited by A. Dold and B. Eckmann

971

Kleinian Groups and Related Topics

Proceedings of the Workshop
Held at Oaxtepec, Mexico A

Edited by D. M. Gallo and R. M. Porter

Springer-Verlag
Berlin Heidelberg New York 1983

Editors

Daniel M. Gallo
R. Michael Porter
Departamento de Matemáticas
Centro de Investigación y de Estudios Avanzados del I.P.N.
Mexico City, Mexico

AMS Subject Classifications (1980): 30F XX, 30F40, 32G15, 51M10, 14H35

ISBN 3-540-11975-2 Springer-Verlag Berlin Heidelberg New York
ISBN 0-387-11975-2 Springer-Verlag New York Heidelberg Berlin

Library of Congress Cataloging in Publication Data. Main entry under title: Kleinian groups and related topics. (Lecture notes in mathematics; 971) Papers presented at the Workshop on Kleinian Groups and Related Topics held in Oaxtepec, Mexico, Aug. 10–14, 1981, during the Second Coloquio de Matemáticas held by the Mathematics Dept. of the Centro de Investigación y de Estudios Avanzados del Instituto Politécnico Nacional. 1. Kleinian groups–Congresses. 2. Riemann surfaces–Congresses. 3. Geometry, Hyperbolic–Congresses. 4. Geometry, Algebraic–Congresses. I. Gallo, D. M. (Daniel M.), 1944-. II. Porter, R. M. (R. Michael), 1952-. III. Workshop on Kleinian Groups and Related Topics (1981: Oaxtepec, Mexico) IV. Series: Lecture notes in mathematics (Springer-Verlag) ; 971.
QA3.L28 no. 971 [QA331] 510s [515'.223] 82-19656
ISBN 0-387-11975-2 (U.S.)

Printing and binding: Beltz Offsetdruck, Hemsbach/Bergstr.
2146/3140-543210

INTRODUCTION

The Workshop on Kleinian Groups and Related Topics was held in Oaxtepec, Mexico, from August 10 to 14, 1981, during the second Coloquio de Matemáticas held by the Mathematics Department of the Centro de Investigación y de Estudios Avanzados del Instituto Politécnico Nacional.

The theory of Kleinian groups has undergone a vast diversification in the last two decades, particularly with its more recent applications to 3-manifold theory. The object of this conference was to provide a stimulus to its development and to make recent progress more accesible to researchers in Mexico. The keynote speaker was B. Maskit.

The contributions to this volume were provided by participants at the Workshop as well as others who responded to a call for papers. In accordance with an agreement made at the time of the conference, all articles contained here have been refereed.

The editors wish to express their appreciation to the Consejo Nacional de Ciencia y Tecnología, the Secretaría de Educación Pública, the Instituto Politécnico Nacional, and the Centro de Investigación y de Estudios Avanzados for their support of this event. In addition we thank all those who had a share in the production of this book, including the Workshop participants, the referees, and Springer-Verlag.

Daniel M. Gallo

R. Michael Porter

TABLE OF CONTENTS

LIFTING SURFACE GROUPS TO SL(2,$\mathbb{C}$)*

W. ABIKOFF

K. APPEL

P. SCHUPP

It is an elementary consequence of the uniformization theorem that the fundamental group $\pi_1 S$ of a closed orientable surface S of genus $g \geq 2$ has a faithful representation as a Fuchsian group G. That the totality of such representations forms a connected subset of the real Möbius group, $\text{Möb}_{\mathbf{R}}$, was first shown by Fricke. Since $\text{Möb}_{\mathbf{R}}$ is canonically isomorphic to $PSL(2,\mathbf{R})$, it is natural and computationally relevant to investigate whether a given representation may be lifted to $SL(2,\mathbf{R})$. Fricke showed that it is indeed possible and further that the lift depends continuously on the surface S viewed as a point in Teichmüller space. It is an immediate consequence of the definitions that given a Möbius transformation γ, there are two choices of lifts $\hat{\gamma}$ of γ to $SL(2,\mathbf{R})$. Since G is a surface group of rank $2g$ there are $2g$ choices of the signs of the generators $\hat{\gamma}_i$. Further, the commutator relation

$$\prod_{i \text{ odd}} [\gamma_i, \gamma_{i+1}] = \text{id}$$

lifts to the relation

$$R = \prod_{i \text{ odd}} [\hat{\gamma}_i, \hat{\gamma}_{i+1}] = \pm \text{Id} \tag{1}$$

where the γ_i are a standard set of generators for G. Fricke's argument shows that there is a constant choice of sign possible in Equation 1. Siegel [5] was first to raise the question as to whether the sign was always positive. Both Bers [3] and Abikoff [2, p. 18] claimed that the result was true. Bers' unpublished proof consists

* This research has been partially supported by the National Science Foundation.

of studying the side identifications of regular $4g$-gons in the hyperbolic plane.

Here we prove the result by showing the following

THEOREM. *If* $\hat{G}$ *is a lift of the surface group* G *to* $SL(2,\mathbf{R})$ *and the genus of* G *satisfies* $g \geq 2$, *then for any choice* $\gamma_1,\ldots,\gamma_{2g}$ *of standard generators of* G, *we have*

$$\prod_{i \text{ odd}} [\hat{\gamma}_i, \hat{\gamma}_{i+1}] = \text{Id}.$$

That this proof appear in print is in response to a question raised by Irwin Kra and the members of a student seminar at Stony Brook.

The proof of the theorem is an immediate consequence of the following three lemmas and the techniques used in the study of augmented Teichmüller spaces (see Abikoff [1] and [2] and Harvey [4]).

If H is a Fuchsian group representing a thrice punctured sphere then G is a free group on two generators but is usually presented as $< \gamma_1,\gamma_2,\gamma_3 : \gamma_1\gamma_2\gamma_3 = \text{id} >$. It is easy to see that by appropriate choice of sign of $\text{tr}\,\hat{\gamma}_3$, the relation may be chosen to lift to

$$\hat{\gamma}_1\hat{\gamma}_2\hat{\gamma}_3 = \text{Id}. \tag{2}$$

LEMMA 1. *If* H *is lifted to* $SL(2,\mathbf{R})$ *so that the traces* $\text{tr}\,\hat{\gamma}_1$ *and* $\text{tr}\,\hat{\gamma}_2$ *have the same sign and the relation* (2) *is valid, then* $\text{tr}\,\hat{\gamma}_3$ *has negative sign.*

LEMMA 2. *If*

$$\hat{\theta} = \prod_{\substack{i \text{ odd} \\ i < 2g-1}} [\hat{\gamma}_i, \hat{\gamma}_{i+1}],$$

then $\text{tr}\,\hat{\theta}$ *and* $\text{tr}\,R$ *have opposite sign.*

LEMMA 3. $\operatorname{tr} \hat{\theta} < 0$.

Lemmas 2 and 3 immediately give the desired result.

1. Proof of Lemma 1.

We normalize $\hat{\gamma}_1$ and $\hat{\gamma}_2$ as follows:

$$\hat{\gamma}_1 = \begin{bmatrix} \pm 1 & \pm 1 \\ 0 & \pm 1 \end{bmatrix} \quad \text{and} \quad \hat{\gamma}_2 = \begin{bmatrix} \pm 1 & 0 \\ \pm\alpha & \pm 1 \end{bmatrix} .$$

Since γ_3^{-1} is parabolic, $\operatorname{tr} \gamma_3^{-1} = \pm 2$, but $\operatorname{tr} \gamma_3^{-1} = 2 + \alpha$. The choice of 2 for $\operatorname{tr} \gamma_3^{-1}$ yields $\alpha = 0$ which is impossible since H is a free group of rank 2.

2. Proof of Lemma 2.

Since $R = \pm \mathrm{Id}$ we have

$$\hat{\theta}\,\hat{\gamma}_{2g-1} = \pm \hat{\gamma}_{2g} \hat{\gamma}_{2g-1} \hat{\gamma}_{2g}^{-1} . \tag{3}$$

Using Fricke's result and the connectedness of the augmented Teichmüller space, we see that a consistent choice of sign may be made for any discrete faithful representation of $\pi_1 S$ in $SL(2,\mathbb{R})$ and that the choice must persist as θ becomes parabolic. We may then assume, by conjugation, that

$$\hat{\theta} = \begin{bmatrix} \pm 1 & \pm 1 \\ 0 & \pm 1 \end{bmatrix} \quad \text{and} \quad \hat{\gamma}_{2g-1} = \begin{bmatrix} a & b \\ c & d \end{bmatrix} .$$

Since trace is a conjugation invariant, Equation 3 implies

$$\operatorname{tr} \hat{\theta}\hat{\gamma}_{2g-1} = \pm \operatorname{tr} \hat{\gamma}_{2g-1}$$

or

$$\pm(a + c + d) = \pm(a + d).$$

If both choices of sign are identical, then $c = 0$. It follows that the Möbius transformation $\gamma_{2g-1} : z \mapsto (az+b)/(cz+d)$ fixes infinity. Since θ also fixes infinity, well-known properties of Fuchsian groups show that γ_{2g-1} and θ commute, but group theory says that they don't.

3. *Proof of Lemma 3.*

In the augmented Teichmüller space a thrice punctured sphere is a limiting deformation of a torus with one hole. Two of the punctures come from pinching one curve; the corresponding matrices are then conjugate in the limit group, hence have traces of the same sign. It follows from Lemma 1 that the sign of the matrix representing the remaining puncture, which came from the border curve, is negative.

If S_g is a surface of genus g, we dissect it as shown in Figure 1

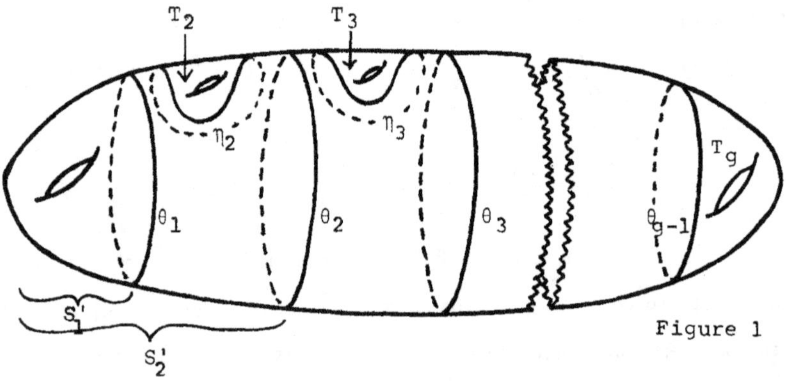

Figure 1

and use induction to show that $\operatorname{tr}\hat{\theta} = \operatorname{tr}\hat{\theta}_{g-1} < 0$. For $g = 2$, it was done above. For $g > 2$, S_g is the union of a surface S'_{g-2} of genus $g-2$ with one border curve θ_{g-2}, a pants and two tori T_{g-1} and T_g each missing a disk. S_g is also the union of S'_{g-1} and T_g. As inductive hypothesis we take $\operatorname{tr}\hat{\theta}_{g-2} < 0$. Since η_{g-1} borders T_{g-1}, $\operatorname{tr}\hat{\eta}_{g-1} < 0$. By Lemma 2, it follows that $\operatorname{tr}\hat{\theta}_{g-1} < 0$.

4. *Some Remarks.*

After proving this result, we learned that several other proofs exist. These other proofs use heavy machinery which is either topological, algebraic or cohomological in nature.

Andrew Haas has observed that this proof may be used to show that $\mathrm{tr}\,\hat{\theta} < 0$ whenever θ is a simple dividing loop on a compact surface.

DEPARTMENT OF MATHEMATICS
UNIVERSITY OF ILLINOIS AT URBANA-CHAMPAIGN
URBANA, ILLINOIS 61801

REFERENCES

[1] W. Abikoff, Degenerating families of Riemann surfaces, *Ann. of Math.* 105 (1977), 29-44.

[2] ——, *The Real Analytic Theory of Teichmüller Space*, Springer Lecture Notes in Mathematics 820 (1980).

[3] L. Bers, Spaces of Riemann surfaces, *Proceedings of the 1958 International Congress of Mathematicians*, 349-361.

[4] W.J. Harvey, Spaces of discrete groups, *Discrete Groups and Automorphic Functions*, Proceedings of the NATO Instructional Conference, Academic Press, London 1977, 295-348.

[5] C.L. Siegel, Über einige Ungleichungen bei Bewegungsgruppen in der nichteuklidischen Ebene, *Math. Ann.* 113 (1957), 127-138.

NEC GROUPS AND KLEIN SURFACES

EMILIO BUJALANCE

1. *Introduction*

In the last few years several results on non-Euclidean crystallographic groups and on applications of these groups to the study of Klein surfaces have been obtained. This paper consists mainly of a brief survey of these results.

In Section 2 NEC groups are introduced and their main properties are given. In Section 3 we study the NEC normal subgroups of NEC groups. In Section 4 we show the existing relations between NEC groups and Klein surfaces as well as between normal NEC subgroups and automorphism groups of Klein surfaces. Finally in Section 5 we study the automorphism groups of Klein surfaces.

2. *NEC groups*

Let us consider transformations of the Riemann sphere $\mathbb{C}^+$ of the following forms:

i) $w(z) = \dfrac{az+b}{cz+d}$, $z \in \mathbb{C}^+$: $ad - bc = 1$, $a,b,c,d \in \mathbb{R}$;

ii) $w(z) = \dfrac{a\bar{z}+b}{c\bar{z}+d}$, $z \in \mathbb{C}^+$: $ad - bc = -1$, $a,b,c,d \in \mathbb{R}$.

These transformations form a group G which maps the upper half plane $\operatorname{Im} z > 0$, which we denote by D, into itself.

The transformations of the form i) preserve the orientation of D and form a subgroup of G of index two (*the hyperbolic group*); the transformations of the form ii) reverse the orientation of D.

The transformations of the hyperbolic group are of three types:

1) *hyperbolic*: if $|a+d| > 2$ with two fixed points on the real axis,
2) *elliptic*: if $|a+d| < 2$ with one fixed point in D, and
3) *parabolic*: if $|a+d| = 2$ with one fixed point of multiplicity two on the real axis.

The transformations of the form ii) are either *glide-reflections* (if $a+d \neq 0$) with two fixed points on the real axis, or *reflections* (if $a+d=0$) with a circle of fixed points.

If, on D, we introduce the Riemannian metric $ds = |dz|y^{-1}$ $(z = x+iy)$, then D becomes a model of the hyperbolic plane and G its group of isometries.

A *non-Euclidean crystallographic* (NEC) group is a discrete subgroup Γ of G with compact quotient space D/Γ. An NEC group containing only orientation-preserving transformations is a *Fuchsian group*.

In order to derive presentations for NEC groups, Wilkie [17] proved the following result:

Let Γ be an NEC group. Then there is a fundamental region P for Γ which is a polygon in D whose perimeter, described counterclockwise, is one of the following:

(1) $\xi_1\xi_1'\cdots\xi_\tau\xi_\tau' \quad \varepsilon_1\gamma_{10}\cdots\gamma_{1s_1}\varepsilon_1' \;\cdots\; \varepsilon_k\gamma_{k0}\cdots\gamma_{ks_k}\varepsilon_k' \quad \alpha_1\beta_1\alpha_1'\beta_1'\cdots\alpha_g\beta_g\alpha_g'\beta_g'$

(2) $\xi_1\xi_1'\cdots\xi_\tau\xi_\tau' \quad \varepsilon_1\gamma_{10}\cdots\gamma_{1s_1}\varepsilon_1' \;\cdots\; \varepsilon_k\gamma_{k0}\cdots\gamma_{ks_k}\varepsilon_k' \quad \alpha_1\alpha_1^*\cdots\alpha_g\alpha_g^*$

In these symbols, each letter denotes an oriented side of the polygon. The apostrophe means that the corresponding sides $\xi\xi', \alpha\alpha', \beta\beta', \varepsilon\varepsilon'$ of the polygon are identified by generators of the group which preserve orientation, and the asterisk means that the corresponding sides $\alpha\alpha^*$ of the polygon are identified by generators of the group which reverse orientation. As a consequence, if we identify corresponding points on the related edges of the polygon, we obtain from it a surface with boundary. In the case (1), the surface will be a sphere with k disks removed and g handles added. In the case (2) the surface will be a sphere with k disks removed and g cross-caps added.

Moreover, we have the following properties: a) the stabilizer in Γ of the vertex M_i common to the sides ξ_i, ξ_i' is a cyclic group of rotations of order m_i (x_i will denote a generator of the group); b) the stabilizer of the vertex N_{ij} common to the sides $\gamma_{i(j-1)}$, γ_{ij} is a dihedral group of order $2n_{ij}$, with generators $c_{i(j-1)}$, c_{ij}, that are reflections in $\gamma_{i(j-1)}$, γ_{ij} respectively; c) the stabilizer of an inner point of the side γ_{ij} joining N_{ij} and $N_{i(j+1)}$ is a group isomorphic to $\mathbb{Z}_2$, generated by c_{ij}; d) the stabilizer of the vertex $N_{i(s_i+1)}$ common to the sides γ_{is_i} and ε_i' is a group isomorphic to $\mathbb{Z}_2$, generated by c_{is_i}. No other points of P are fixed points for Γ.

Macbeath [9] associated to the NEC groups of type (1) the NEC *signature:*

$$(*) \quad (g\,;+\,;[m_1,\ldots,m_\tau]\,;\ \{(n_{11}\ldots n_{1s_1}),\ldots,(n_{k1}\ldots n_{ks_k})\})$$

and to the groups of type (2) the signature

$$(**) \quad (g\,;-\,;[m_1,\ldots,m_\tau]\,;\ \{(n_{11}\ldots n_{1s_1}),\ldots,(n_{k1}\ldots n_{ks_k})\})$$

These signatures characterize the algebraic structure of the groups.

Wilkie [17] proves that an NEC group with signature (*) has the presentation given by generators:

$x_i : i = 1,\ldots,\tau$	elliptic transformations
$a_j, b_j : j = 1,\ldots,g$	hyperbolic transformations
$c_{ij} : i = 1,\ldots,k;\ j = 0,\ldots,s_i$	reflections
$e_i : i = 1,\ldots,k$	hyperbolic transformations

and relations:

$$x_i^{m_i} = 1 : \ i = 1,\ldots,\tau$$

$$c_{is_i} = e_i^{-1} c_{i0} e_i : \ i = 1,\ldots,k$$

$$c_{i(j-1)}^2 = c_{ij}^2 = (c_{i(j-1)}c_{ij})^{n_{ij}} = 1 : \ i = 1,\ldots,k\,;\ j = 1,\ldots,s_i$$

$$x_1 \dots x_\tau e_1 \dots e_k a_1 b_1 a_1^{-1} b_1^{-1} \dots a_g b_g a_g^{-1} b_g^{-1} = 1.$$

In a group Γ with signature (**) we have the presentation given by generators:

$x_i : \ i = 1,\dots,\tau$	elliptic transformations
$d_j : \ j = 1,\dots,g$	hyperbolic transformations
$c_{ij} : \ i = 1,\dots,k ; \ j = 0,\dots,s_i$	reflections
$e_i : \ i = 1,\dots,k$	hyperbolic transformations

and relations:

$$x_i^{m_i} = 1 : \ i = 1,\dots,\tau$$

$$c_{is_i} = e_i^{-1} c_{i0} e_i : \ i = 1,\dots,k$$

$$c_{i(j-1)}^2 = c_{ij}^2 = (c_{i(j-1)} c_{ij})^{n_{ij}} = 1 : \ i = 1,\dots,k ; \ j = 1,\dots,s_i$$

$$x_1 \dots x_\tau e_1 \dots e_k d_1^2 \dots d_g^2 = 1.$$

From now on, we will denote by $x_i, e_i, c_{ij}, a_i, b_i, d_j$ the above generators associated with an NEC group.

Let c, c' be two period cycles $c = (n_1 \dots n_s)$ and $c' = (n_1' \dots n_s')$. Then c, c' are called *directly equivalent* if one is a cyclic permutation of the other, that is, if $s = s'$ and there is an integer k such that $n_i = n'_{k+i}$, suffixes being read modulo s.

c, c' are called *inversely equivalent* if one is a cyclic permutation of the other reversed, that is, if $s = s'$ and there is an integer k such that $n_i = n'_{k-i}$, where the suffixes are again reduced modulo s.

If Γ is a NEC group of signature

$$(g ; \pm ; [m_1 \dots m_\tau] ; \{(c_1 \dots c_k)\})$$

then Γ is isomorphic to the group Γ' of signature

$$(g' ; \pm ; [m'_1 \ldots m'_{\tau'}] ; \{(c'_1 \ldots c'_{k'})\})$$

if and only if they have the same sign, $g = g'$, $\tau = \tau'$, $k = k'$, $[m'_1, \ldots m'_{\tau'}] = [m_1 \ldots m_\tau]$, and there is a permutation ϕ of $(1,2\ldots k)$ such that either (a) for each i, c'_i is directly equivalent to $c_{\phi(i)}$, or (b) for each i, c'_i is inversely equivalent to $c_{\phi(i)}$, in the case of the sign "+" ; and there is a permutation of $(1\ldots k)$ such that for each i, c'_i is either directly or inversely equivalent to $c_{\phi(i)}$ in the case of the sign "-" (Macbeath [9] and Wilkie [17]).

In the same way as in the case of Fuchsian groups, in the NEC groups the measures of the fundamental regions will allow us to relate the NEC signatures of the groups with those of the subgroups. Similarly these measures give some conditions for the achievement of the NEC signatures. We will call the *measure of a fundamental region of an NEC group* Γ the area of any fundamental region associated to the group, and we will denote it by $|\Gamma|$. The measure of the fundamental region of an NEC group with signature (*) is (Singerman [16])

$$|\Gamma| = \pi\left(2\tau + 2k + 4g - 4 + \sum_{i=1}^{k} s_i - \sum_{i=1}^{k}\sum_{j=1}^{s_i} \frac{1}{n_{ij}} - \sum_{i=1}^{\tau} \frac{2}{m_i}\right) = \pi R$$

and the measure of the fundamental region of an NEC group Γ with signature (**) is

$$|\Gamma| = \pi\left(2\tau + 2k + 2g - 4 + \sum_{i=1}^{k} s_i - \sum_{i=1}^{k}\sum_{j=1}^{s_i} \frac{1}{n_{ij}} - \sum_{i=1}^{\tau} \frac{2}{m_i}\right) = \pi R'$$

If the expressions R and R' are > 0, then there are NEC groups of signatures (*) and (**) respectively (Zieschang [18]).

3. *Normal subgroups of NEC groups*

In this section we will describe some results that have been obtained for NEC normal subgroups of an NEC group.

If Γ is an NEC group and Γ' is a normal subgroup of Γ with finite index, then Γ' is an NEC group. Conversely, every NEC normal subgroup of an NEC group has finite index.

The orientation in the signatures of the NEC groups is the same for the normal subgroups in the case that the index of the group with respect to the subgroup is odd (Bujalance [2], Hall [6], Hoare, Singerman [7]).

If Γ is a *proper NEC group* (i.e., non Fuchsian), then it has a subgroup Γ^+ of index 2 consisting of the elements which preserve orientation (i.e., $\Gamma^+ = \Gamma \cap G^+$) called the *canonical Fuchsian group* of Γ. In [16] Singerman determines the signature of Γ^+ given that of Γ.

In [2], [3] we have studied the existing relations between the signatures of NEC normal subgroups of an NEC group and the signature of the group, when the index of the group is a prime number $p \neq 2$. As a consequence we have obtained necessary and sufficient criteria on the NEC signatures for the existence of an NEC group with such a signature, which is a normal subgroup of some other NEC group with index different from a power of 2.

We have also established necessary and sufficient conditions for a normal subgroup of odd index of an NEC group to be of a specific signature. These conditions have been obtained using the existence of a certain group of permutations. We have calculated the proper periods of normal NEC subgroups when the index of the subgroup is even.

4. *Compact Klein surfaces*

By a *Klein surface* we shall mean a surface X with or without boundary together with an open covering by a family of sets $\mathcal{U} = \{U_i\}$ with the following properties:

1) For each $U_i \in \mathcal{U}$ there exists a homeomorphism φ_i of U_i onto an open set in the sphere $\mathbb{C}^+$.

2) If $U_i, U_j \in \mathcal{U}$ and $U_i \cap U_j \neq \emptyset$ then $\varphi_i \circ \varphi_j^{-1}$ is an analytic or antianalytic mapping defined on $\varphi_j(U_i \cap U_j)$.

A homeomorphism $f: X \longrightarrow X$ is a called an *automorphism* if $\varphi_i \circ f \circ \varphi_j^{-1}$ is either an analytic or antianalytic mapping in its domain of definition.

If X is orientable and without boundary we will say that it is a *Riemann surface*.

Now let E be the field of all meromorphic functions of a Klein surface X. E is an algebraic function field in one variable over $\mathbb{R}$ and as such it has an algebraic genus. In case X is a Riemann surface, the algebraic genus is equal to the topological genus of X. If X is not a Riemann surface, then the relationship between the topological genus p and the algebraic genus g is given by

$p = (g - k + 1)/2$ if X is orientable,

$p = g - k + 1$ if X is non-orientable,

where k denotes the number of boundary components of X.

We will say that an NEC group Γ is a *surface group* if the quotient space $X = D/\Gamma$ is compact and the quotient map $p : D \longrightarrow D/\Gamma$ is unramified. Γ will be called a *bordered surface* if X has boundary.

The following results give the relationship between Klein surfaces and NEC groups:

If Γ is an NEC group, then the quotient space D/Γ has a unique dianalytic structure such that the quotient map $P : D \longrightarrow D/\Gamma$ is a morphism of Klein surfaces (Alling and Greenleaf [1]).

Let X be a compact Klein surface of algebraic genus $g \geq 2$. Then X can be represented in the form D/Γ, where Γ is a bordered surface group or a surface group (Singerman [15] and Preston [14]).

We close this section by describing the relation between the automorphism groups of Klein surfaces and the NEC normal subgroups of an NEC group.

In the case of compact non-orientable Klein surfaces without boundary, a necessary and sufficient condition for a finite group G to be a group of automorphisms of a non-orientable Klein surface without boundary of topological genus $p \geq 3$ is that there exist a proper NEC group Γ and a homomorphism $\theta : \Gamma \longrightarrow G$ such that the kernel of θ is a surface group and $\theta(\Gamma^{+}) = G$ (Singerman [15]).

In the case of compact Klein surfaces with boundary, we have that if Γ is a bordered surface group, G is a group of automorphisms of the Klein surface D/Γ if and only if $G = \Gamma'/\Gamma$, where Γ' is an NEC group such that $\Gamma \subset \Gamma' \subset N(\Gamma)$ (May [10]).

5. *Automorphisms of compact Klein surfaces*

In this section, we give some results on automorphism groups of compact Klein surfaces that are analogous to the ones obtained in the last few years for Riemann surfaces.

Singerman in [15] has established that the bound given by Hurwitz [8] for the order of a group of automorphisms of a Riemann surface works for compact non-orientable Klein surfaces without boundary of algebraic genus $g \geqslant 2$, and he has found families of compact non-orientable Klein surfaces without boundary for which the bound is attained.

May in [10], [12] and [13] has shown that if $\upsilon(g)$ is the order of the largest group of automorphisms which a compact Klein surface with non empty boundary of algebraic genus $g \geqslant 2$ can admit, then $12(g-1) \geqslant \upsilon(g) \geqslant 4(g+1)$. In both cases, he also has found infinitely many values of g for which these bounds are attained.

These upper bounds may be considered as particular cases of the general problem of finding the minimum genus of surfaces for which a given finite group G is a group of automorphisms. This problem, for the case of cyclic groups of automorphisms of compact non-orientable Klein surfaces without boundary, has been solved by Bujalance [5]. In the case of groups of automorphisms of compact Klein surfaces with one boundary component it has been solved by Bujalance [4].

The problem of the greatest order of an automorphism of a Klein surface has been solved for the case of non-orientable Klein surfaces without boundary by Bujalance [5] and Hall [6], and for the case of compact Klein surfaces with boundary by May [11].

DEPARTAMENTO DE MATEMATICA FUNDAMENTAL, FACULTAD DE CIENCIAS MATEMATICAS
UNIVERSIDAD NACIONAL DE EDUCACION A DISTANCIA (SPAIN)

REFERENCES

[1] N.L. Alling and N. Greenleaf, *Foundations of the theory of Klein surfaces*, Lecture Notes in Math. No. 219, Springer-Verlag (1971).

[2] E. Bujalance, Normal subgroups of NEC groups, *Math. Zeit.* 173 (1981), 331-341.

[3] __________, Proper periods of normal NEC groups with even index, to appear in *Revista Hispano-Americana*.

[4] __________, Automorphism groups of compact Klein surfaces with one boundary component, to appear.

[5] __________, Cyclic groups of automorphisms of compact non-orientable Klein surfaces without boundary, to appear.

[6] W. Hall, Automorphisms and coverings of Klein surfaces, Ph.D. Thesis, Southampton (1978).

[7] A.H.M. Hoare and D. Singerman, The orientability of subgroups of plane groups, to appear.

[8] A. Hurwitz, Über algebraische Gebilde mit eindeutigen transformationen in sich, *Math. Ann.* 41 (1893), 428-471.

[9] A.M. Macbeath, The clasification of non-Euclidean plane crystallographic groups, *Can. J. Math.* 19 (1967), 1192-1205.

[10] C.L. May, Large automorphism groups of compact Klein surfaces with boundary-I, *Glasgow Math. J.* 18 (1977), 1-10.

[11] _______, Cyclic automorphism groups of compact bordered Klein surfaces, *Houston J. Math.*, 3 (1977), 395-405.

[12] _______, Automorphisms of compact Klein surfaces with boundary, *Pacific J. Math.* 59 (1975), 199-210.

[13] _______, A bound for the number of automorphisms of a compact Klein surface with boundary, *Proc. Amer. Math. Soc.* 63 (1977), 273-280.

[14] R. Preston, Projective structures and fundamental domains on compact Klein surfaces, Ph.D. Thesis, University of Texas (1975).

[15] D. Singerman, Automorphisms of compact non-orientable Riemann surfaces, *Glasgow J. Math.* (1971), 50-59.

[16] __________, On the structure of non-Euclidean crystallographic groups, *Proc. Cam. Phil. Soc.* (1974), 223-240.

[17] C.H. Wilkie, On non-euclidean crystallographic groups, *Math. Zeit.* 91 (1966), 87-102.

[18] H. Zieschang, E. Voght and H.D. Coldewey, *Flächen und ebene diskontinuierliche Gruppen*, Lecture Notes in Math. No. 122, Springer-Verlag (1970); *Surfaces and Planar Discontinuous Groups*, Lecture Notes in Math. No. 835, Springer-Verlag (1980).

REMARKS ON THE CYCLOTOMIC FRICKE GROUPS

HARVEY COHN

This note presents in some detail two special classical Fuchsian groups of Fricke's type. They are called "cyclotomic" because of symmetries in their fundamental domains; i.e., the modules are those of the equilateral triangle and the square, corresponding to the third and fourth roots of unity as the case may be. They have appeared in the literature on previous occasions in seemingly unrelated contexts. The first is from the standard (Klein) modular group $SL(2,\mathbb{Z})$, (see, e.g., [10]) and the second is from the so-called "Hecke group for $\sqrt{2}$" (see, e.g., [8]). In a series of papers, (some of which [4], [5] have been presented at previous conferences in this series), the author showed how the first synthesizes such topics as

Markoff's minimal quadratic forms
primitive elements in a free group of rank two
geodesics of a punctured torus.

More recently, Asmus Schmidt presented a general approach to this synthesis (see [12]) which drew further attention to the analogies of these special cases.

While we should not hesitate to declare such examples to be "sui generis", a conceptualization of their common features will aid in approaching unresolved problems involving trace estimates and involving the generalization process to the Hilbert modular group on the product of half-planes (which is of only recent scrutiny, see [9]).

We now start with the groups of Klein and Hecke, both called G, in matrix form within $SL(2,\mathbb{R})$, and in linear-fractional (or projective) form within $PSL(2,\mathbb{R})$ $(= SL(2,\mathbb{R})/(\pm I))$. (Here I is the identity matrix and 1 is the projective identity $\pm I$).

Klein's Group G

$$R = \begin{pmatrix} 1 & 1 \\ 0 & 1 \end{pmatrix}$$

$$S = \begin{pmatrix} 0 & -1 \\ 1 & 0 \end{pmatrix}$$

$$Q = RS = \begin{pmatrix} 1 & -1 \\ 1 & 0 \end{pmatrix}$$

$$S^2 = Q^3 = -I$$

Hecke's Group G

$$R = \begin{pmatrix} 1 & \sqrt{2} \\ 0 & 1 \end{pmatrix}$$

$$S = \begin{pmatrix} 0 & -1 \\ 1 & 0 \end{pmatrix}$$

$$Q = RS = \begin{pmatrix} \sqrt{2} & -1 \\ 1 & 0 \end{pmatrix}$$

$$Q^2 = \begin{pmatrix} 1 & -\sqrt{2} \\ \sqrt{2} & -1 \end{pmatrix}$$

$$S^2 = Q^4 = -I$$

Projective forms of either group G are as follows:

$$G = (S,Q;\ S^2 = Q^3 = 1)$$

$$G = (S,Q;\ S^2 = Q^4 = 1)$$

Fundamental domains of G for the upper half z-plane are shown in the figure below.

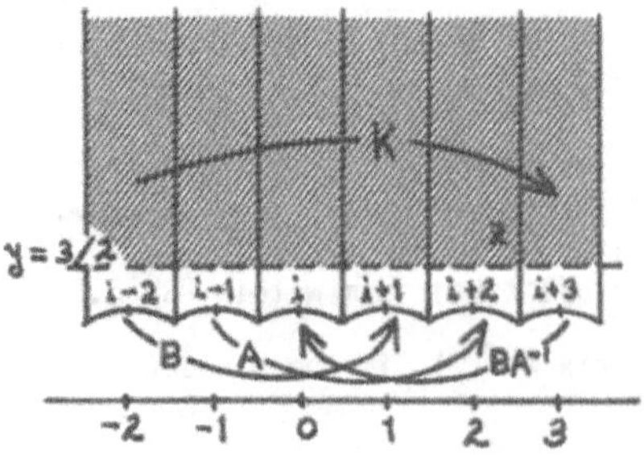

Figure 1a. Klein's group. (six replicas)

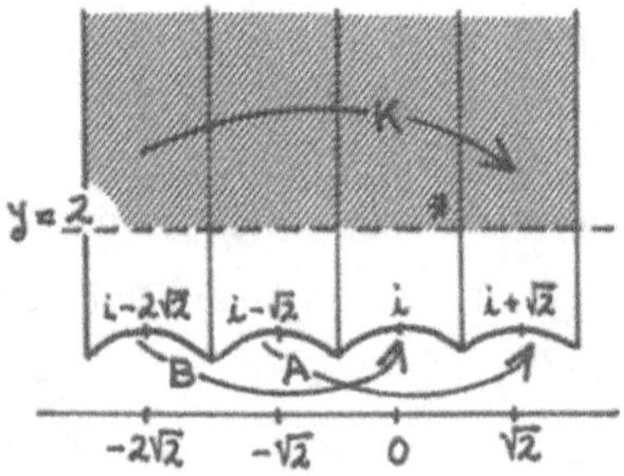

Figure 2a. Hecke's group. (four replicas)

Commutator-related subgroups (note $[X,Y] = XYX^{-1}Y^{-1}$) are

$$A = [Q,S] = \begin{pmatrix} 2 & 1 \\ 1 & 1 \end{pmatrix}$$

$$B = [Q^{-1},S] = \begin{pmatrix} 1 & 1 \\ 1 & 2 \end{pmatrix}$$

$$BA^{-1} = \begin{pmatrix} 0 & 1 \\ -1 & 3 \end{pmatrix}$$

$$K = [A,B^{-1}] = -\begin{pmatrix} 1 & 6 \\ 0 & 1 \end{pmatrix}$$

$$A^* = AB^{-1}A = \begin{pmatrix} 5 & 2 \\ 2 & 1 \end{pmatrix}$$

$G' = [G,G] = (A,B) = G^*$

$G/G' = \{Q^t S^u\}$

(t mod 3; u mod 2)

$[G:G'] = 6$

$$A = Q^2S^{-1} = \begin{pmatrix} \sqrt{2} & 1 \\ 1 & \sqrt{2} \end{pmatrix}$$

$$B = SQSQS = \begin{pmatrix} 0 & -1 \\ 1 & 2\sqrt{2} \end{pmatrix}$$

$$K = [A,B^{-1}] = -\begin{pmatrix} 1 & 4\sqrt{2} \\ 0 & 1 \end{pmatrix}$$

$$A^* = AB^{-1} = \begin{pmatrix} 3 & \sqrt{2} \\ \sqrt{2} & 1 \end{pmatrix}$$

$(A^* = [Q,S^{-1}])$

$G^* = (A,B) = (A,A^*)$

$[Q^2,S] = A^2$

$[Q^{-1},S] = AB$, $[S,Q] = BA^{-1}$

$G' = [G,G] = (A^2,AB,B^2) \subseteq G^*$

$G/G' = \{Q^t S^u\}$

(t mod 4; u mod 2)

$[G:G'] = 8$

$G/G^* = \{Q^t\}$

$[G:G^*] = 4$, $[G^*:G'] = 2$

he fundamental domain of G^* (torus) is drawn in the following figure.

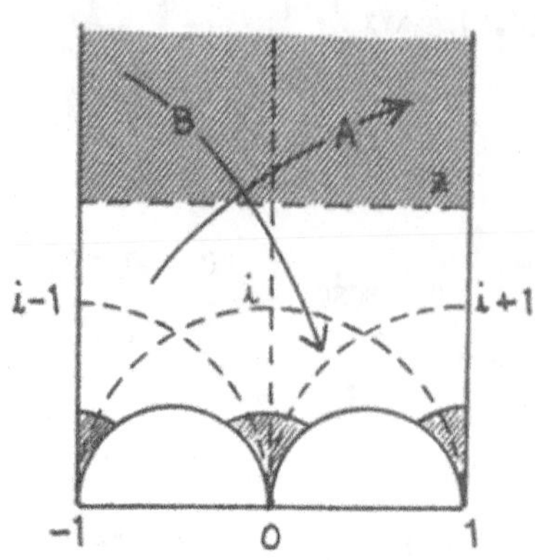

Figure 1b.

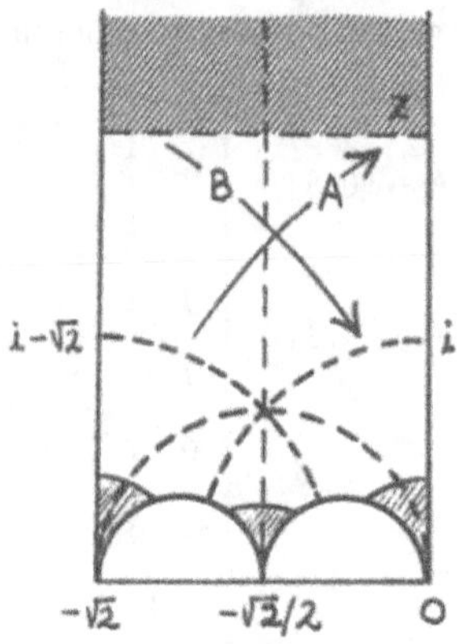

Figure 2b. (Note G' is the subgroup of G* consisting of words of even length in A and B .)

For either Klein's or Hecke's group G, we now arrive at

$$G^* = (A,B),$$

a Fricke group. It is characterized as a free group with two hyperbolic generators and a parabolic commutator. It is of genus one. When its generators are taken as matrices in $SL(2,\mathbb{R})$, we can set

$$\text{trace } A = a, \quad \text{trace } B = b, \quad \text{trace } AB^{-1} = c$$

and we can compute traces directly from Fricke's relations

(M)
$$a^2 + b^2 + c^2 = abc$$
$$\text{trace } A^{-1}B^{-1} = ab - c$$

Specifically, we use the above operation with its symmetrical counterparts, starting from

$$(a,b,c) = (3,3,3) \qquad \Bigg| \qquad (a,b,c) = (\sqrt{8},\sqrt{8},4)$$

We next make use of a theorem of the author on the elements of a free group (of order two) which are primitive (i.e., any such can be an element of a free basis). We consider first the so-called "step word" given by the formula (see [2])

$$M(u,v) = (A,B)^{u,v} = \prod_{s=1}^{u} AB^{[su/v]-[(s-1)u/v]}$$

for $u>0$. As special cases (or definitions),

$$(A,B)^{0,0} = 1, \quad (A,B)^{0,1} = B, \quad (A,B)^{1,0} = A. \quad \text{etc.}$$

For $u<0$ we define $M(u,v)$ by $M(-u,-v)M(u,v) = 1$. To see that $M(u,v)$ is primitive when $(u,v) = 1$, we need only note that

$$(A,B) = (M(u,v),M(u',v'))$$

if $uv'-vu' = 1$. Also, $M(u,v)^n = M(nu,nv)$.

Now, take an arbitrary word W in G^*. It abelianizes to the form $uA+vB$. If W is primitive, then $(u,v) = 1$, since the abelianized word must be primitive in $\mathbb{Z} \oplus \mathbb{Z}$. Conversely, the theorem asserts that W is primitive only if it is equivalent modulo commutators to $M(u,v)$ for $(u,v) = 1$. In general, if $k = (u,v) > 0$, we can write any element of W, easily, as

$$W = XM(u/k,v/k)^k \qquad (= X \quad \text{if} \quad k=0)$$

for X an element in the commutator group $[G^*,G^*]$, (which is characterized by zero abelianization).

The Poincaré invariant metric, $ds = |dz|/y$, leads to geodesics in the usual fashion, (semicircles with diameter on the real axis). Of these, the closed geodesics are, easily, those which connect real "conjugate" fixed points of the hyperbolic matrices in G. It is, however, a significant matter that the closed geodesics which lie below

y = 3/2 | y = 2

belong to matrices equivalent to one of the set M(u,v), (see [2] and [12]). Their traces form a (so-called) *Markoff triple* (satisfying (M), see above). Note that closed geodesics below the above bounds are merely closed geodesics on a perforated torus, a noncompact surface of negative curvature, shown by the shaded areas above. Since trace is essentially the length of the geodesic or an eigenvalue in a spectrum, depending on the context, the question of trace is of major interest.

For many purposes, accordingly, we may wish to consider only the set of traces, and therefore we wish to eliminate repetitions through symmetries. For example, in the above examples, if A and B were interchanged the values of M would be unchanged. We now consider the set of such symmetries.

This is best seen by reference to the fundamental domain for the abelianization of the group G* seen by taking the U-plane where U is the abelian integral of the first kind. The U-plane is a period parallelogram generated from J(z), the usual modular invariant which maps the fundamental domain of G in the z-plane onto the J-plane so that the boundary goes into the real axis. Then we refer to

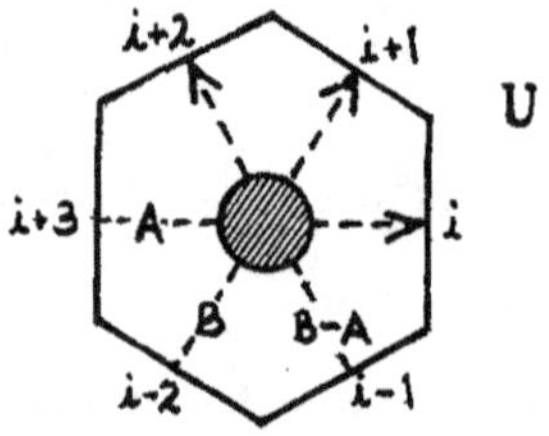

Figure 1c.

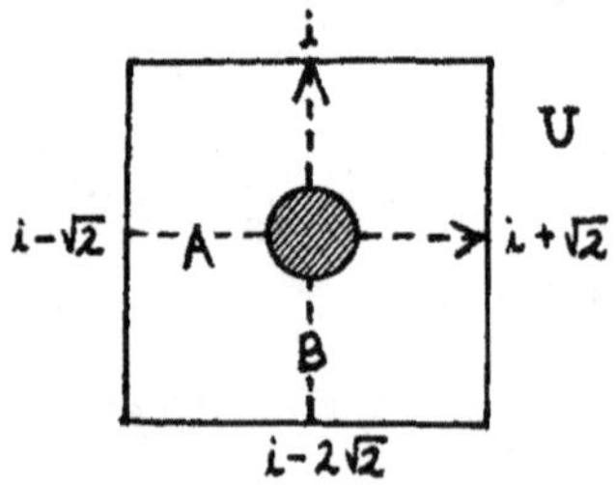

Figure 2c.

with U defined by

$dU = \text{const}\ \dfrac{dJ}{J^{2/3}(J-1)^{1/2}}$	$dU = \text{const}\ \dfrac{dJ}{J^{1/2}(J-1)^{1/2}}$

From the symmetry of the U-parallelogram, we can say $M(u,v) = M(u',v')$ if $(u.v)$ and (u',v') represent vectors equivalent under a symmetry group D which is dihedral

of order 12.	of order 8.

Such symmetries are represented projectively in terms of

$$t = u/v, \qquad t' = u'/v'$$

by the transformations $t \mapsto t'$ where t' has the values

$\{t,1-t,1/t,(t-1)/t,t/(t-1),1/(1-t)\}$.	$\{t,-t,1/t,-1/t\}$.

The full dihedral group D is realized by a semidirect product with the transformation $(u,v) \mapsto (-u,-v)$ which does not change the value of t. We write out the group in terms of new generators which are more convenient for trace calculations,

$$G^* = (A^*,A), \qquad M(u,v) = (A^*,A)^{u,v}.$$

The dihedral groups D have as generators these matrices acting on the exponents u and v:

reflections

$\begin{pmatrix} u \\ v \end{pmatrix} \mapsto \begin{pmatrix} 1 & 1 \\ 0 & -1 \end{pmatrix} \begin{pmatrix} u \\ v \end{pmatrix}$	$\begin{pmatrix} u \\ v \end{pmatrix} \mapsto \begin{pmatrix} 1 & 0 \\ 0 & -1 \end{pmatrix} \begin{pmatrix} u \\ v \end{pmatrix}$

rotations

of order 6	of order 4
$\begin{pmatrix} u \\ v \end{pmatrix} \mapsto \begin{pmatrix} 2 & 1 \\ -3 & -1 \end{pmatrix} \begin{pmatrix} u \\ v \end{pmatrix}$	$\begin{pmatrix} u \\ v \end{pmatrix} \mapsto \begin{pmatrix} 1 & 1 \\ -2 & -1 \end{pmatrix} \begin{pmatrix} u \\ v \end{pmatrix}$

CONJECTURE A

For primitive elements (relatively prime u,v) which are not necessarily non-negative, we can have

$$|\text{trace } M(u,v)| = |\text{trace } M(u',v')|$$

if and only if (u,v) and (u',v') are connected by the symmetry group D. Thus, by this conjecture, all traces of primitive elements are given uniquely by "first quadrant" values:

$$\{\text{trace } M(u,v) : u \geq 0,\ v \geq 0,\ (u,v) = 1\}.$$

CONJECTURE B

If the word W in G^* is abelianized to $k(uA^* + vA)$ with $(u,v) = 1$ and (say) $k > 0$, then the trace in the matrix representation has the estimate

$$|\text{trace } W| \geq |\text{trace } M(u,v)^k|.$$

Thus, the minimum length of the geodesic in each homotopy class is that of the homologous power of a primitive element.

These two conjectures represent a peculiar situation since they seem obvious from "geometric intuition" and have withstood extensive testing yet have not been proved. Conjecture A is (in Klein's case) attributed to Dickson and Cassels (see [6] and [1]), but the originating reference seems elusive. It is purely diophantine in nature since it refers to equation (M) whose solutions are integers or integers times $\sqrt{2}$ in the respective cases. Conjecture A asserts, independently of the theory of Fuchsian groups, that each triple (a,b,c) satisfying (M) is uniquely determined by its maximum entry (also see [11]). Likewise Conjecture B, made by the author [5], belongs to matrix theory (or possibly to the calculus of variations in the large). It is hoped that this survey will present them in the proper context to encourage further investigation. It would be no surprise if the tools required actually now exist (see [13]).

CITY COLLEGE (CUNY)
NEW YORK, NY 10031

REFERENCES

[1] J.W.S. Cassels, "The Markov chain," *Ann. of Math.* 50 (1949), 676-685.

[2] H. Cohn, "Representation of Markoff's binary quadratic forms by geodesics on a perforated torus," *Acta. Arith.* 18 (1971), 125-136.

[3] H. Cohn, "Markoff forms and primitive words," *Math. Ann.* 196 (1972), 8-22.

[4] H. Cohn, "Some direct limits of primitive homotopy words and of Markoff geodesics," Conf. Disc. Groups and Riemann Surf., *Annals of Math. Studies* 79, Princeton (1974), 81-98.

[5] H. Cohn, "Minimal geodesics on Fricke's torus-covering," in *Riemann Surfaces and Related Topics*, Proc. 1978 Stony Brook Conference, Princeton (1980), 73-85.

[6] L.E. Dickson, *Studies in the theory of numbers*, Chicago (1930), 79-107.

[7] R. Fricke, "Über die Theorie der automorphen Modulgruppen," *Gott. Nach.* (1896), 91-101.

[8] E. Hecke, "Über die Bestimmung Dirichletscher Reihen durch ihre Funktionalgleichung," *Math. Ann.* 112 (1936), 664-699.

[9] F. Kirchheimer and J. Wolfart, "Explizite Präsentation gewisser Hilbertscher Modulgruppen durch Erzeugende und Relationen," *J. reine angew. Math.* 315 (1980), 139-173.

[10] F. Klein and R. Fricke, *Vorlesungen uber die Theorie der Elliptischen Modulfunktionen*, Leipzig (1890).

[11] A.A. Markoff, "Sur les formes binaires indéfinies, I," *Math. Ann.* 15 (1879), 391-409; II, *Math. Ann.* 17 (1890), 379-400.

[12] A.L. Schmidt, "Minimum of quadratic forms with respect to Fuchsian groups, I," *J. reine angev. Math.* 286/287 (1976), 341-348.

[13] H. Zieschang, E. Vogt, and H.D. Coldewey, "Flächen und ebene diskontinuerliche Gruppen," Lecture Notes 122, Springer Verlag, Berlin and New York (1970).

ON THE NOETHER GAP THEOREM

MICHAEL ENGBER

ABSTRACT

Some of the facts immediately surrounding the Weierstrass Gap Theorem are not true when restated for the Noether Gap Theorem. For example, the sum of two Noether non-gaps need not be a non-gap. On the other hand, like Weierstrass points, Noether sequences form an analytic set of codimension 1. In fact, both can be described as the zero loci of appropriate differentials. In addition, the gap sequences on a hyperelliptic surface are of special type and, for arbitrary surfaces, a "principle of non-degeneracy" holds.

1. Let X be a compact Riemann surface of genus $g > 1$ and let $P_1, P_2, \ldots$ be a sequence of points of X. Let $D_k = P_1 + \ldots + P_k$ and $D_0 = 0$. In analogy with the considerations that lead to the Weierstrass Gap Theorem, we ask:

(*) For each nonnegative k, does there exist a meromorphic function f on X whose polar divisor $(f)_\infty$ is precisely D_k?

This turns out, however, not to be quite the right family of questions. The correct question is the following:

(**) For each nonnegative k, does there exist a meromorphic function f on X whose polar divisor satisfies $(f)_\infty \leq D_k$, $(f)_\infty \nleq D_{k-1}$?

If the answer to (**) for a given k is no, then we say that k is a (Noether) gap for the sequence $P_1, \ldots$ Otherwise, k is a non-gap.

THEOREM 1 (Noether Gap Theorem). *With the above notation, there are precisely* g *gaps and all gaps are* $\leq 2g-1$.

Proof. See for example [2, Theorem III.5.4, p. 79].

In the light of this theorem, it is clear that we need consider only sequences of points of length 2g-1. If all these points coincide, then the questions (*) and (**) coincide and Theorem 1 is the Weierstrass Gap Theorem. In the general case, an affirmative answer for given k to (*) implies an affirmative answer to (**) but not conversely.

DEFINITION 1. Given a sequence $P_1, \ldots, P_{2g-1}$ on X, we say that k is an *exact* non-gap if and only if there exists a meromorphic function f on X such that $(f)_\infty = D_k$.

As in the context of the Weierstrass Gap Theorem, we define the weight of a sequence.

DEFINITION 2. Given a sequence $P_1, \ldots, P_{2g-1}$ on X, let $r_1, \ldots, r_g$ be the gaps. The *weight*, ν, of the sequence is given by:

$$\nu = \sum_{i=1}^{g} r_i - g(g+1)/2$$

THEOREM 2. Let $P_1, \ldots, P_{2g-1}$ *be points on* X. *Assume the* P_i *are ordered so that* ν *is a maximum. Then every non-gap is exact.*

LEMMA 1. Let n *and* $n-1$ *be non-gaps. If* $n-1$ *is exact then so is* n.

Proof. By hypothesis, there exist functions f and g such that $(g)_\infty = D_{n-1}$ and $(f)_\infty = P_n + D'$, where $D' \leq D_{n-1}$. But then, for almost all $c \in \mathbb{C}$, $(f+cg)_\infty = D_n$.

Proof of Theorem 2. First note that 0 is always an exact non-gap for any sequence (since a constant has no poles). Let n be the smallest non-exact non-gap. Then there exists an f such that $P_n < (f)_\infty < D_n$. (The left inequality is strict because there are no functions of order 1. The right inequality is strict because n is not exact.) Let m be the largest non-gap less than n. Since m is necessarily exact, the lemma implies that $m < n-1$, whence $m+1, m+2, \ldots, n-1$ is a non-void sequence of gaps. Since there exists a g such that $(g)_\infty = D_m$, there exists $c \in \mathbb{C}$ such that

$f_1 = f + cg$ has polar divisor greater than $D_m + P_n$. Since n is not exact, $(f_1)_\infty < D_n$ and therefore there exists r with $m < r < n$ such that $(f_1)_\infty \leqslant D_n - P_r$. Since $m+1, \ldots, n-1$ is a sequence of gaps, it follows that no permutation of $P_{m+1}, \ldots, P_{n-1}$ will change the overall gap sequence nor, *a fortiori*, the weight ν. Thus we may assume that $r = n-1$.

Now consider the sequence $P_1, \ldots, P_{n-2}, P_n, P_{n-1}, P_{n+1}, \ldots, P_{2g-1}$. Let D'_k be the sum of the first k points in this sequence. Then $D'_k = D_k$ unless $k = n-1$. Therefore, no gaps or non-gaps have changed except that $n-1$ is now a non-gap and n is now a gap. Thus ν has been increased by 1, contradicting maximality.

2. By analogy with the notion of Weierstrass point, we define:

DEFINITION 3. A *Noether sequence* is a sequence $P_1, \ldots, P_{2g-1}$ such that it or a permutation of it has positive weight.

Let X^n (resp. $X^{(n)}$) denote the cartesian (resp. symmetric) product of n copies of X and let $b_n : X^n \longrightarrow X^{(n)}$ be the canonical projection.

PROPOSITION 1. The set of Noether sequences in X^{2g-1} *is the set*

$$b_{2g-1}^{-1}(b_{2g-1}(a^{-1}(b_g^{-1}(G_g^2))))$$

where $a : X^{2g-1} \longrightarrow X^g$ *is the projection onto the first* g *factors and* G_g^2 *is the set of effective divisors,* D, *of degree* g *with* $\ell(D) \geqslant 2$.

Recall that $\ell(D)$ is the dimension of the $\mathbb{C}$-vector space of meromorphic functions f such that $(f) + D \geqslant 0$.

Proof. To say that a divisor of degree g has dimension $\geqslant 2$ is to say that there exists a non-constant function with poles among the points comprising it. Thus any sequence involving these g points will in some order have a non-gap $\leqslant g$.

Note: The set G_g^2 may also be described as the zero locus of the unique (up to multiplication by a constant) holomorphic g-form on $X^{(g)}$. (See [1, Prop. 6, p. 812].) This characterization is analogous to that of the set of Weierstrass points as the zeroes of the "Wronskian" differential on X.

THEOREM 3. The set of Noether sequences is of codimension 1 *in* X^{2g-1}.

Proof. G_g^2 is of codimension 1 in $X^{(g)}$ by [3, Theorems 7, 14b, 15]. Since b_n is a finite map, both it and its inverse preserve dimension (and codimension). Since a is the projection of a cartesian product, a^{-1} preserves codimension.

3. *Hyperelliptic surfaces*

In this section, X will be assumed hyperelliptic. $\varphi: X \longrightarrow \mathbb{P}^1$ is a 2-sheeted cover and $\sigma: X \longrightarrow X$ is the sheet interchange.

DEFINITION 4. A sequence of points $P_1, \ldots, P_n$ is *σ-paired up to* 2r if and only if $P_{2i} = \sigma(P_{2i-1})$ for $i = 1, \ldots, r$.

The following proposition generalizes a well-known result.

PROPOSITION 2. Let f *be a meromorphic function on* X *with polar divisor* $(f)_\infty = P_1 + \ldots + P_n$ *and suppose* $P_1, \ldots, P_n$ *is σ-paired up to* $2r$. *If* $n = \operatorname{ord} f \leq g + r$, *then* n *is even and* f *is a rational function of* φ.

Proof. Induction on r. If $r = 0$, the proposition is well-known. See, for example [2, Proposition III.7.10, p. 102].

Let $r \geq 1$. By hypothesis, $(f)_\infty = P_1 + \sigma(P_1) + \ldots + P_r + \sigma(P_r) + D$ where $D \geq 0$. Let Q be any zero of f and write $(f)_0 = Q + Z$. ($(f)_0$ denotes the divisor of zeroes of f.) By composing φ, if necessary, with a linear fractional transformation on $\mathbb{P}^1$, we may assume that $(\varphi) = Q + \sigma(Q) - P_r - \sigma(P_r)$. But then

$$(f/\varphi) = (f) - (\varphi)$$

$$= Z - D - (P_1 + \sigma(P_1) + \ldots + P_{r-1} + \sigma(P_{r-1})) - \sigma(Q).$$

Now, $\mathrm{ord}(f/\varphi) \leq n-1 \leq g+r-1$. Therefore, by the induction hypothesis, $\mathrm{ord}(f/\varphi)$ is even and f/φ is a rational function of φ. Since $\mathrm{ord}\,\varphi$ is even the result follows.

Note that f is a rational function of φ if and only if $f = f \circ \sigma$.

THEOREM 4. *Let* $P_1, \ldots, P_{2g-1}$ *be a sequence on* X *such that every non-gap is exact. Then there exists a non-negative integer* $s \leq g-1$ *such that the gap sequence consists of all integers* $1, \ldots, g+s$ *other than* $2, 4, \ldots, 2s$. *Furthermore, all such gap sequences occur.*

Proof. If there are no even gaps then $s = g-1$. Assume then that there exist even gaps. Anticipating the conclusion, let $2s+2$ denote the smallest even gap. Then $2s+2 \leq 2g-1$ whence $s \leq g-2$. We will first show that $P_1, \ldots, P_{2g-1}$ is σ-paired up to $2s$. Since $2,4,\ldots,2s$ are exact non-gaps, there exists a meromorphic function f_i such that $(f_i)_\infty = P_1 + \ldots + P_{2i}$ for each $i = 1,\ldots,s$. We now proceed by induction.

If $s = 1$, then $\mathrm{ord}\, f_1 = 2 \leq g$ so that $f_1 = f_1 \circ \sigma$ whence $(f_1)_\infty = \sigma(f_1)_\infty$ so that $P_2 = \sigma(P_1)$. Suppose $s > 1$. Then by the induction hypothesis, the sequence is σ-paired up to $2s-2$. Since $\mathrm{ord}\, f_s = 2s$ and $s \leq g-2$, $\mathrm{ord}\, f_s \leq g+s-1$ and we may apply Proposition 2 with $r = s-1$. Thus $f_s = f_s \circ \sigma$, whence $P_{2s} = \sigma(P_{2s-1})$ so that the sequence is σ-paired up to $2s$. Again by the proposition, all odd numbers $\leq g+s$ are gaps.

Now let k be the smallest non-gap with $2s+2 < k \leq g+s$. Since k is exact, there exists h with $(h)_\infty = P_1 + \ldots + P_k$. Since the P_i are σ-paired up to $2s$ and $\mathrm{ord}\, h = k \leq g+s$, k is even and $h = h \circ \sigma$, so that σ permutes $P_1, \ldots, P_k$. We already know that $P_{2i} = \sigma(P_{2i-1})$ for $i \leq s$. Therefore, $\sigma(P_{2s+1}) = P_\ell$ where

$2s+2 \leq \ell \leq k$. If $\ell = 2s+2$ then it is easy to find a function with polar divisor D_{2s+2}, but then $2s+2$ is not a gap, contrary to our assumption. Thus $\ell > 2s+2$. Let m be the integer (necessarily $\leq k$) such that $P_m = \sigma(P_{2s+2})$. (It is possible that P_{2s+1} (resp. P_{2s+2}) is a fixed point of σ. In such a case, there must nonetheless be another point among $P_{2s+3}, \ldots, P_k$ which is equal to P_{2s+1} (resp. P_{2s+2}). It is this other point that is to be denoted by P_ℓ (resp. P_m).) Suppose $\ell < m$. Then $D = P_1 + \ldots + P_{2s+1} + D + \sigma(P_{2s+1})$ with $D \geq 0$ and it is again easy to find a function in $L(D_\ell)$ which is not in $L(D_{\ell-1})$. Thus ℓ is a non-gap, contrary to the minimality of k. A similar argument applies if $m < \ell$. Thus all k in the interval $2s+2 \leq k \leq g+s$ are gaps. Since we already know that all odd numbers less than or equal to $g+s$ are gaps, we have accounted for all g gaps.

Conversely, given any $s \leq g-1$, choose $P_1, \ldots, P_s$ so that $\sigma(P_i) \neq P_j$ for $i \neq j$ and choose $Q_1, \ldots, Q_{2g-2s-1}$ so that $\sigma(Q_i) \neq Q_j$ and $\sigma(P_i) \neq Q_j$ for all i and j. Then the sequence

$$P_1, \sigma(P_1), \ldots, P_s, \sigma(P_s), Q_1, \ldots, Q_{2g-2s-1}$$

has the desired gap sequence.

4. *Principle of Non-degeneracy*

In [5, p. 559], Rauch enunciates the principle of non-degeneracy which may be formulated as follows: Let $T_{g,1}$ denote the Teichmüller space of Riemann surfaces of genus g with a distinguished point, then the set of points in $T_{g,1}$ at which the first Weierstrass non-gap is $\leq k$ is closed.

If $p: V \to T$ is the universal family of Riemann surfaces of genus g over Teichmüller space, let $F_n(V/T)$ denote the n-fold fiber product of V with itself over T. Then $F_n(V/T)$ parametrizes n-tuples of points on Riemann surfaces of genus g.

The analogue of Rauch's theorem is

THEOREM 5. For any k, *let* $N(k) \subseteq F_{2g-1}(V/T)$ *be the set of* 2g-1-*tuples with first (Noether) non-gap less than or equal to* k. *Then* N(k) *is a closed analytic set.*

Proof. Note that if $k > g$, the assertion is trivial since $N(k) = F_{2g-1}$. Let

$$N'(k) = \{(t, P_1, \dots, P_k) \in F_k(V/T) \mid \text{there exists a non-gap} \leq k\}.$$

Meis [4, Satz 30, p. 37] shows that $N'(k)$ is a closed analytic set. Let $a : F_{2g-1}(V/T) \longrightarrow F_k(V/T)$ be the projection onto the first k factors. Then $N(k) = a^{-1}(N'(k))$. Since a is holomorphic the result follows.

DEPARTMENT OF MATHEMATICS
CUNY
NEW YORK, NY 10031

BIBLIOGRAPHY

[1] A. Andreotti, "On a theorem of Torelli," *Amer. J. of Math.*, 80 (1958), 801-828.

[2] H.M. Farkas and I. Kra, *Riemann Surfaces*, Springer, New York, Heidelberg, Berlin, 1980.

[3] R.C. Gunning, *Lectures on Riemann surfaces, Jacobi varieties*, Princeton University Press, Princeton, N.J., 1972.

[4] Th. Meis, "Die minimale Blätterzahl der Konkretisierungen einer kompakten Riemannschen Fläche," *Schriftenreihe des Math. Inst.*, Münster, 1960.

[5] H.E. Rauch, "Weierstrass points, branch points and moduli of Riemann surfaces," *Comm. Pure Appl. Math.*, 12 (1959), 543-560.

A 3-DIMENSIONAL HYPERBOLIC COLLAR LEMMA

DANIEL GALLO

1. *Introduction*

We consider a discrete subgroup Γ of $PSL(2,\mathbb{C})$ acting in the standard manner on $\mathbb{H}^3 = \{(z,t);\ z \in \mathbb{C},\ t > 0\}$, the upper half space model of hyperbolic space. If $X \in \Gamma - \{id.\}$ is a non-parabolic element, we denote by ℓ_X the geodesic in $\mathbb{H}^3$ joining the fixed points of X. For $k \geqslant 0$ we define $B_k(\ell_X) = \{p \in \mathbb{H}^3;\ d(p,\ell_X) \leqslant k\}$ where d is the hyperbolic metric. Let $<X>^I$ be the subgroup of Γ which leaves ℓ_X invariant. $B_k(\ell_X)$ is called a *collar for* X if

$$Y(B_k(\ell_X)) \cap B_k(\ell_X) = \emptyset \quad \text{for all} \quad Y \in \Gamma - <X>^I \quad \text{and}$$

$$Y(B_k(\ell_X)) = B_k(\ell_X) \quad \text{for all} \quad Y \in <X>^I .$$

In this note we show that there exists a universal constant $\varepsilon_0 > 0$ (i.e. independent of Γ) and a function $k : (0,\varepsilon_0) \longrightarrow \mathbb{R}^+$ such that all elements $X \in \Gamma$ have a collar $B_{k(\varepsilon)}(\ell_X)$ whenever $0 < |tr^2(X) - 4| \leqslant 4\varepsilon < 4\varepsilon_0$. Moreover, for two such elements these collars are disjoint. We note that $\lim_{\varepsilon \to 0} k(\varepsilon) = \infty$. The proof follows from Jørgensen's inequality (see [4]).

Our result is a corollary to Proposition 1 which gives a lower bound for the real hyperbolic distance between the axes of two elements of small complex translation length in a group Γ. We remark that corresponding versions of Proposition 1 and Corollaries 1 and 2 may be obtained for elements of small real translation length. Full details will appear elsewhere.

The referee has pointed out that an analogue of Corollary 1 has recently appeared in an article by R. Brooks and P. Matelski [2]. The constants we obtain here are better than those in [2]. In particular we are able to show the existence of disjoint collars for elliptic elements of order $\geqslant 8$. We also note an application of Proposition 1 to torsion free groups with quotients of finite volume.

The author wishes to thank his colleague Michael Porter for a crucial observation which led to a considerable improvement of the constants involved in Corollary 1.

We assume the reader is familiar with the terminology of discrete subgroups of $PSL(2,\mathbb{C})$ and refer him to [1] for precise definitions.

2. *Results*

Suppose that $X = \begin{pmatrix} a & b \\ c & d \end{pmatrix} \in PSL(2,\mathbb{C})$ (where $ad-bc = 1$) is not parabolic and does not have 0 or ∞ as a fixed point in $\partial\mathbb{H}^+ \cup \{\infty\}$. Let B_X be the complex distance[1] from ℓ_X to the geodesic joining 0 to ∞ and let τ_X be the complex translation length of X. We begin with the formula

$$|\cosh(B_X)| = \frac{|a-d|}{2|\sinh(\tau_X/2)|} . \tag{1}$$

This may easily be verified using the euqations of [1] for example.

PROPOSITION 1. Let $k > 0$. *Then there exists* $\varepsilon > 0$ *with the following property: Let* $\Gamma \subset PSL(2,\mathbb{C})$ *be a discrete group and suppose* $X_i \in \Gamma$ *satisfy* $0 < |tr^2(X_i)-4| \leqslant \varepsilon$, $i = 1,2$. *Then* $d(\ell_{X_1}, \ell_{X_2}) \geqslant k$ *whenever* $\ell_{X_1} \neq \ell_{X_2}$.

Proof. Choose $\varepsilon > 0$ such that $0 < 4|\sinh(\tau/2)|^2 \leqslant \varepsilon$ implies

$$\frac{1}{|\sinh(\tau/2)|^2} - 4(1 + |\sinh(\tau/2)|^2) \geqslant \cosh^2(k) \tag{2}$$

Now assume that

$$0 < |tr^2(X_2) - 4| \leqslant |tr^2(X_1) - 4| \leqslant \varepsilon \tag{3}$$

---

1. We will only be concerned with $|\cosh(B_X)|$ so that there is no ambiguity arising from the orientation of ℓ_X (see [1]).

and normalize Γ so that $X_1 = \begin{pmatrix} \sqrt{\gamma} & 0 \\ 0 & 1/\sqrt{\gamma} \end{pmatrix}$, $X_2 = \begin{pmatrix} a & b \\ c & d \end{pmatrix}$, where $e^{\tau_{X_1}} = \gamma$ and $ad-bc = 1$. Since $\varepsilon < 1$, X_2 is not elliptic of order two and we may apply Jørgensen's inequality

$$|\mathrm{tr}^2(X_1) - 4| + |\mathrm{tr}(X_1X_2X_1^{-1}X_2^{-1}) - 2| \geq 1$$

which we write

$$4|\sinh(\tau_{X_1}/2)|^2 + 4|ad-1||\sinh(\tau_{X_1}/2)|^2 \geq 1.$$

Thus

(4) $$4|ad-1| \geq \frac{1}{|\sinh(\tau_{X_1}/2)|^2} - 4 .$$

Using the equality

$$(a-d)^2 = (a+d)^2 - 4 - 4(ad-1)$$

and inequalities (2), (3) and (4) we obtain

(5) $$\begin{aligned} |a-d|^2 &\geq 4|ad-1| - 4|\sinh(\tau_{X_2}/2)|^2 \\ &\geq \frac{1}{|\sinh(\tau_{X_1}/2)|^2} - 4 - 4|\sinh(\tau_{X_2}/2)|^2 \\ &\geq \cosh^2(k). \end{aligned}$$

We now assume $\ell_{X_1} \neq \ell_{X_2}$. In order to apply (1) we need to show that X_1 and X_2 do not share a fixed point in $\partial\mathbb{H}^+ \cup \{\infty\}$. This is clear if either X_1 or X_2 is loxodromic. If both elements are elliptic and share a common fixed point in

$\partial\mathbb{H}^+ \cup \{\infty\}$ then they must belong to a finite extension of a parabolic subgroup of Γ (see [3]). In this case the only possible orders for X_1 and X_2 are 2, 3, 4 and 6 which contradicts $|tr^2(X_i) - 4| < 1$, $i = 1,2$. Let B be the complex distance between ℓ_{X_1} and ℓ_{X_2} (note that ReB is the real hyperbolic distance). Now (1) and (5) yield

$$|\cosh(B)| = \frac{|a-d|}{2|\sinh(\tau_{X_2}/2)|} \geq \cosh(k).$$

We then have $\cosh(\mathrm{Re}B) \geq \cosh(k)$ and $\mathrm{Re}B \geq k$.

COROLLARY 1 (Collar Lemma). *There exists an* ε_0, $1 > \varepsilon_0 > 0$, *with the following property: Let* $\Gamma \subset PSL(2,\mathbb{C})$ *be a discrete group and let* $X \in \Gamma$ *be such that* $0 < |\sinh(\tau_X/2)|^2 \leq \varepsilon < \varepsilon_0$. *Then* ℓ_X *has a collar* $B_{k(\varepsilon)}(\ell_X)$ *(where* $k(\varepsilon)$ *depends only on* ε*). Moreover, if* $0 < |\sinh(\tau_{X_i}/2)|^2 \leq \varepsilon < \varepsilon_0$ *for* $X_i \in \Gamma$, $i = 1,2$, *then* $B_{k(\varepsilon)}(\ell_{X_1}) \cap B_{k(\varepsilon)}(\ell_{X_2}) = \emptyset$ *whenever* $\ell_{X_1} \neq \ell_{X_2}$.

Proof. Choose the largest $\varepsilon_0 > 0$ such that $0 < |\sinh(\tau/2)|^2 \leq \varepsilon_0$ implies

$$(6) \qquad \frac{1}{|\sinh(\tau/2)|^2} - 4(1 + |\sinh(\tau/2)|^2) \geq 1.$$

Let $0 < |\sinh(\tau_X/2)|^2 \leq \varepsilon < \varepsilon_0$. Since $tr^2(X) = tr^2(Y^{-1}XY)$ for all $Y \in \Gamma$, it follows as in the proof of Proposition 1 that the distance between ℓ_X and any of its translates under $Y \in \Gamma - \langle X \rangle^I$ is $\geq \operatorname{arc\,cosh}\left(\frac{1}{\varepsilon^2} - 4(1+\varepsilon^2)\right) = 2k(\varepsilon)$. $B_{k(\varepsilon)}(\ell_X)$ is the required collar.

Solving for ε_0 in (6) one obtains $\varepsilon_0 = \frac{\sqrt{41} - 5}{8}$ (compare with [2] where the constant ε_0 obtained satisfies $\varepsilon_0 = \frac{1}{8}$). Thus if X and Y are elliptic of order ≥ 8 (and their axes are distinct) we choose X' and Y' in the subgroups generated by X and Y respectively, such that $|tr^2(X') - 4|$, $|tr^2(Y') - 4| \leq 4\sin^2(\pi/8) < 4\varepsilon_0$. Therefore X and Y have disjoint collars of length $k(\sin^2(\pi/8))$.

The next corollary applies to discrete torsion-free groups Γ .

COROLLARY 2. Let $c>0$. *Then there exists a constant* $V(c)>0$ *with the following property: If* $X\in\Gamma$ *is loxodromic and* $\mathrm{Vol}(\mathbb{H}^3/\Gamma)\leqslant c$ *then* $|\tau_X|\geqslant V(c)$.

Proof. The volume of the solid torus $T_X = B_{k(\varepsilon)}/<X>^I$ about $\ell_X/<X>^I$ for $X\in\Gamma$ when $0<|\sinh(\tau_X/2)|^2\leqslant\varepsilon<\varepsilon_0$ is given by $\mathrm{Vol}(T_X) = \mathrm{Re}(\tau_X)(\sinh^2(k(\varepsilon)))$. It is clear from (6) that $\lim_{|\tau_X|\to 0}\mathrm{Vol}(T_X) = \infty$ which proves the corollary.

CENTRO DE INVESTIGACION Y DE ESTUDIOS AVANZADOS
MEXICO, D.F., MEXICO

REFERENCES

[1] R. Brooks and J. Matelski, The dynamics of 2-generator subgroups of PSL(2,ℂ), Proceedings of the 1978 Stony Brook Conference, *Ann. of Math. Studies*, 97 (1981), 87-100.

[2] R. Brooks and J. Matelski, Collars in Kleinian groups, *Duke Math. Jour.* 49 (1982), 163-182.

[3] L.R. Ford, *Automorphic Functions*, 2nd. ed., Chelsea Publishing Company, New York, (1951).

[4] T. Jørgensen, On discrete groups of Möbius transformations, *Amer. J. Math.* 98 (1976), 739-749.

PROJECTIVE STRUCTURES ON OPEN SURFACES

DANIEL M. GALLO

R. MICHAEL PORTER

Introduction

The purpose of the present article is to discuss the extension of certain results about projective structures from compact to open Riemann surfaces.

Modern writers who have investigated projective structures include Earle, Gunning, Hejhal, Kra, and Maskit. Most of their attention has focused on the monodromy mapping, in particular, on questions of its local and global injectivity, and related variational formulas.

We are mainly concerned here with bordered surfaces of finite type; for completeness we also briefly discuss punctures. In Section 2 we look at the restriction of the monodromy mapping to the space of bounded quadratic differentials which are real on the boundary curves (for some surfaces it is not injective). In the final section we present a result on the infinite-dimensional space of quadratic differentials having no restriction on their boundary behavior.

1. *Projective structures*

A surface of *finite type* (p,n,m) is obtained by removing m disjoint closed disks and n additional points from a compact Riemann surface of genus p where $p,m,n < \infty$. Let S be a surface of finite type which is represented as the quotient $S = U/G$ where G is a Fuchsian group acting on the upper half plane U. Recall that a projective structure (see [6]) is determined by an analytic cover $\{V_\alpha, w_\alpha\}$, $w_\alpha : V_\alpha \longrightarrow \hat{\mathbb{C}} = \mathbb{C} \cup \{\infty\}$ for which the transition functions are restrictions of elements of the Möbius group $\operatorname{Aut} \hat{\mathbb{C}}$.

A projective structure is most easily studied as a *deformation* (f,χ) of G, where f is a meromorphic local homeomorphism defined in U and $\chi : G \to \operatorname{Aut} \hat{\mathbb{C}}$ is a homomorphism of G such that

$$f \circ \gamma = \chi(\gamma) \circ f \tag{1}$$

for all $\gamma \in G$. The function f yields the coordinates of the projective structure associated to the deformation (f,χ); we call χ the *monodromy homomorphism* associated to the projective structure.

We recall the definitions of the following spaces:

$$F_2(G) = \{\phi : \phi \text{ is holomorphic in } U \text{ and } (\phi \circ \gamma)(\gamma')^2 = \phi \text{ for all } \gamma \in G\} \;;$$

$$B_2(G) = \{\phi \in F_2(G) : \sup_{z \in U} |\phi(z)| (\operatorname{Im} z)^2 < \infty\} \;.$$

Let $\Omega(G)$ be the region of discontinuity of G (note that when $m > 0$, $\Omega(G) \cap \mathbb{R} \neq \emptyset$). Then we define also

$$B_2^*(G) = \{\phi \in B_2(G) \,;\, \phi \text{ has a holomorphic extension to } \Omega(G) \text{ which is real on } \Omega(G) \cap \mathbb{R}\} \;.$$

When $m > 0$, $F_2(G)$ and $B_2(G)$ are infinite dimensional complex vector spaces, while $B_2^*(G)$ is a real vector space of dimension $6p - 6 + 2n + 3m$. An element of $B_2^*(G)$ will be called a *reflectable, bounded quadratic differential.*

It is well known that if (f,χ) is a deformation of G, then its Schwarzian derivative $[f] = (f''/f')' - \frac{1}{2}(f''/f')^2$ lies in $F_2(G)$. We will assume throughout that f is normalized by $f(z_0) = 0$, $f'(z_0) = 1$, $f''(z_0) = 0$, where $z_0 \in U$ is fixed. With this normalization established, each $\phi \in F_2(G)$ determines a unique deformation (f_ϕ, χ_ϕ) (see [6]). Write $\Phi(\phi) = \chi_\phi$ and abbreviate $\mathrm{Hom} = \mathrm{Hom}(G, \mathrm{Aut}\,\hat{\mathbb{C}})$. We have thus defined.

$$\Phi : F_2(G) \longrightarrow \mathrm{Hom},$$

the *monodromy mapping* for G. Hom has a natural differentiable structure induced by the embedding $\chi \mapsto (\chi(\gamma_1),\ldots,\chi(\gamma_r))$ of Hom into $(\mathrm{Aut}\,\hat{\mathbb{C}})^r$, where $\{\gamma_1,\ldots,\gamma_r\}$ is a fixed finite set of generators for G (see [5]). It is easily seen that Φ is holomorphic. Let $\mathrm{Hom}' = \mathrm{Hom}$ modulo inner automorphisms; that is, identify χ with $\tilde{\chi}$ where $\tilde{\chi}(\gamma) = A \circ \chi(\gamma) \circ A^{-1}$, whenever $A \in \mathrm{Aut}\,\hat{\mathbb{C}}$. Let $\pi : \mathrm{Hom} \rightarrow \mathrm{Hom}'$ be the natural projection. Then we also have the mapping $\Phi' = \pi \circ \Phi : F_2 \rightarrow \mathrm{Hom}'$.

2. *Injectivity of the monodromy mapping*

Various proofs of the injectivity of Φ' (and hence of Φ) have been given for the case that $S = U/G$ is a compact surface (see Poincaré [10], Appel-Goursat [1], Gunning [6], Kra [7]). A modern proof is as follows: given $\phi \in F_2(G)$ it is well known that $f_\phi = y^1_\phi / y^2_\phi$ where y^1_ϕ, y^2_ϕ are linearly independent solutions of the linear differential equation $2y'' + \phi y = 0$. Suppose $(f_{\phi_1}, \chi_{\phi_1})$ and $(f_{\phi_2}, \chi_{\phi_2})$ are two deformations whose homomorphisms determine the same element of Hom'. Relaxing the requirement that f_{ϕ_2} be normalized, one may as well assume $\chi_{\phi_1} = \chi_{\phi_2}$. One then observes that $= y^1_{\phi_1} y^2_{\phi_2} - y^1_{\phi_2} y^2_{\phi_1}$ is a multiplicative holomorphic (-1)-differential for G; that is,

$$(\rho \circ \gamma)(\gamma')^{-1} = c_\gamma \rho \tag{2}$$

for all $\gamma \in G$, where $0 \neq c_\gamma \in \mathbb{C}$. Since S is compact, $\rho = 0$ (see [7]) and $f_{\phi_1} = f_{\phi_2}$. We will henceforth write $F_{-1}(G)$ for the collection of holomorphic functions f in U satisfying (2) with $c_\gamma = 1$ for all $\gamma \in G$.

This proof may be extended to show that when S is of type $(p,n,0)$, $n > 0$, the restriction of Φ' to $B_2(G)$ is injective: One writes out ρ in terms of the natural local coordinate of a puncture on S, and checks that it is holomorphic there (see [9]).

When $m > 0$, the restriction of Φ to $B_2^*(G)$ need not be injective (see section 3). Thus if one wants to generalize the injectivity theorem, a somewhat different "monodromy mapping" must be introduced. For simplicty we will restrict the discussion to the case $m = 1$. To phrase Theorem 1A adequately for surfaces with more than one nontrivial boundary component, one would consider a mapping which associates to $\phi \in B_2^*(G)$ not only the homomorphism χ_ϕ but also certain information about the images of the intervals of discontinuity $\Omega(G) \cap \mathbb{R}$ under f_ϕ (see [4]).

Suppose then that $S = U/G$ is a surface of type (p,n,m), $m = 1$, and let $I_0 \subset \mathbb{R}$ be an interval of discontinuity for G. Choose $a_0 \in I_0$ and for $\phi \in B_2^*(G)$ let $A_\phi \in \text{Aut}\,\hat{\mathbb{C}}$ be the unique Möbius transformation such that $g_\phi = A_\phi \circ f_\phi$ satisfies $g_\phi(a_0) = 0$, $g_\phi'(a_0) = 1$, $g_\phi''(a_0) = 0$. (We are using the fact that f_ϕ extends analytically across I_0.) We will denote $A_\phi \circ \chi_\phi \circ A_\phi^{-1}$ by χ_ϕ^*. Define

$$\Phi^* : B_2^*(G) \longrightarrow (\text{Aut}\,\hat{\mathbb{C}})^r$$

by

$$\Phi^*(\phi) = (\chi_\phi^*(\gamma_1),\ldots,\chi_\phi^*(\gamma_r)).$$

THEOREM 1A. Let $S = U/G$ *be a surface of type* $(p,n,1)$, *where* $6p - 3 + 2n > 0$, *and* G *is a Fuchsian group. Then* $\Phi^* : B_2^*(G) \to (\text{Aut}\,\hat{\mathbb{C}})^r$ *is injective.*

We outline a proof here as full details may be found in [4]. Let $S^d = \Omega(G)/G$ be the Schottky double of S. Since $6p - 3 + 2n > 0$ there exists a Fuchsian group G^d acting on U such that $U/G^d = S^d$. Thus there is a holomorphic covering map $h : U \longrightarrow \Omega(G)$ and a homomorphism $\mu : G^d \longrightarrow G$ such that

$$h \circ \gamma = \mu(\gamma) \circ h \tag{3}$$

for all $\gamma \in G^d$ (see [2]).

Let Δ_0 and Δ_0^* be the components of $h^{-1}(U)$ and $h^{-1}(U^*)$ respectively (here U^* is the lower half plane) which are adjacent to a fixed component σ of $h^{-1}(I_0)$. Define $\hat{g}_\phi$ to be the extension of g_ϕ to $U \cup I_0 \cup U^*$ obtained by reflecting g_ϕ across I_0.

Suppose $\phi_1, \phi_2 \in B_2^*(G)$ are such that the corresponding deformations $(g_{\phi_1}, \chi_{\phi_1}^*)$

and $(g_{\phi_2}, \chi^*_{\phi_2})$ share a common homomorphism. Now the quadratic differentials

$$(\phi_1 \circ h)(h')^2 + [h] \quad \text{and} \quad (\phi_2 \circ h)(h')^2 + [h]$$

in $B_2(G^d)$ induce deformations $(f^d_{\phi_1}, \chi^d_{\phi_1})$, $(f^d_{\phi_2}, \chi^d_{\phi_2})$ of G^d where $f^d_{\phi_1} = \hat{g}_{\phi_1} \circ h$, $f^d_{\phi_2} = \hat{g}_{\phi_2} \circ h$ on $\Delta_0 \cup \sigma \cup \Delta^*_0$. It is now easy to verify that $\chi^d_{\phi_1} = \chi^d_{\phi_2}$ on the stabilizers of Δ_0 and Δ^*_0 in G^d. But, since $m = 1$, it follows from van Kempen's theorem that these two stabilizers generate G^d. Thus $\chi^d_{\phi_1} = \chi^d_{\phi_2}$ on all of G^d. We conclude from the results cited earlier that $f^d_{\phi_1} = f^d_{\phi_2}$, from which $\phi_1 = [f_{\phi_1}] = [f_{\phi_2}] = \phi_2$ follows easily.

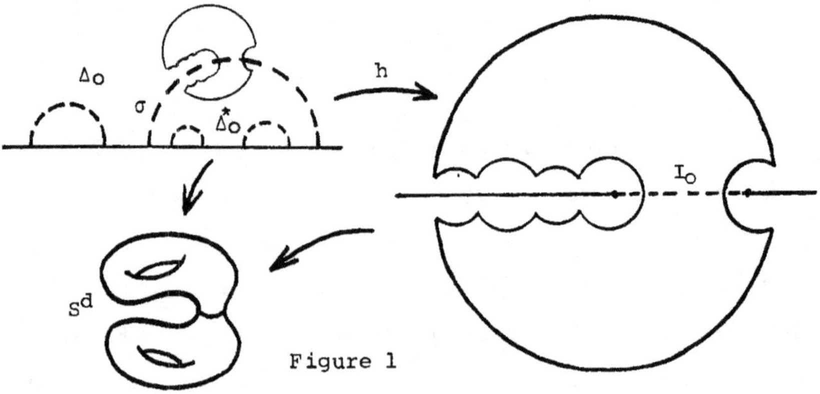

Figure 1

We may construct surfaces S of type $(p,0,m)$, $m \geq 3$ for which Φ is not injective as follows. Let $P(z)$ be a polynomial of degree m whose derivative $P'(z)$ has only simple zeros. Thus as a mapping of $\hat{\mathbb{C}}$, P is branched at precisely m points $z_1 = \infty, z_2, \dots, z_m$ and the values $P(z_j)$, $j = 1,\dots,m$, are distict. Let S_1 be a compact topological surface of genus p and let $h : S_1 \to \hat{\mathbb{C}}$ be a branched covering with branch set contained in $\{q_1, \dots, q_m\}$ such that $h(q_j) = z_j$. (This may be done in many ways, for example when $m = 3$ we may choose h so that it is nontrivially branched only at q_1, q_2, q_3.)

Define

$$S = S_1 - \bigcup_{j=1}^{m} \overline{N_j}$$

where the N_j are neighborhoods of q_j such that $P(h(N_j))$ are disjoint Euclidean discs containing the points z_j respectively. By a classical theorem of conformal

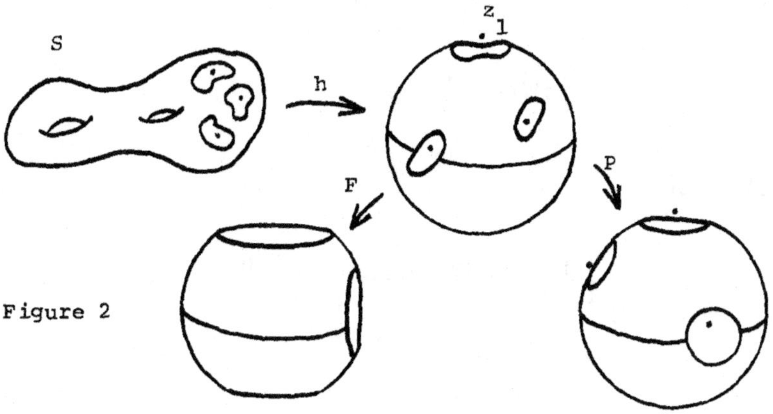

Figure 2

mappings the finitely connected domain $\hat{\mathbb{C}} - \bigcup_{j=1}^{m} \overline{h(N_j)}$ is conformally equivalent, via a univalent mapping F, to a plane domain bounded by m circles.

We now have two topological immersions $F \circ h$ and $P \circ h$ of S into $\hat{\mathbb{C}}$ which induce the same conformal structure on S. We may write $S = U/G$ for some Fuchsian G. The projective structures, however, are distinct since the two mappings do not differ by a Möbius transformation. These projective structures are indeed reflectable since ∂S is sent to a union of circles in each case. The monodromy for each structure is the trivial homomorphism, and therefore $\Phi : B_2^*(G) \to \text{Hom}$ is not injective.

We observe that there cannot exist surfaces of type $(p,0,m)$, $p > 0$, $m = 1,2$, which carry projective structures with a trivial monodromy homomorphism, for the following reason. A reflectable immersion $h : S \to \hat{\mathbb{C}}$ of a surface of type $(p,0,m)$ can be extended to a branched mapping $h : \hat{S} \to \hat{\mathbb{C}}$ where $\hat{S}$ is formed by attaching m discs onto S, the branch points being in the interiors of the attached discs (one branch point for each disc). But there are no mappings from surfaces of genus $p>0$ onto $\hat{\mathbb{C}}$ branched at fewer than three points.

3. *Nonsingularity of the monodromy mapping*

Let S be a compact Riemann surface which is uniformized by a Fuchsian group G acting on U. Fix $\phi_0 \in F_2(G)$ and suppose that the derivative of Φ vanishes in the direction of another differential $\phi \in F_2(G)$; that is,

$$D_{\phi_0}(\Phi)(\phi) = \frac{d}{d\lambda}\Big|_{\lambda=0} \Phi(\phi_0 + \lambda\phi) = 0. \tag{4}$$

From (1) we obtain

$$\left(\frac{df}{dz} \circ \gamma\right) \frac{d\gamma}{dz} = \left(\frac{\partial\chi(\gamma)}{\partial z} \circ f\right) \frac{df}{\partial z} \tag{5a}$$

and

$$\frac{\partial f}{\partial\lambda} \circ \gamma = \left(\frac{\partial\chi(\gamma)}{\partial z} \circ f\right) \frac{\partial f}{\partial\lambda} + \frac{\partial\chi(\gamma)}{\partial\lambda} \circ f \tag{5b}$$

The assumption (4) yields that $\frac{\partial\chi(\gamma)}{\partial\lambda} = 0$ at $\lambda = 0$ for all $\gamma \in G$. Substituting this in (5b) and then dividing (5a) by (5b) we see that the function

$$\rho_f(z) = \left(\frac{\partial f}{\partial\lambda}\right)\left(\frac{\partial f}{\partial z}\right)^{-1}\Big|_{\lambda=0} \tag{6}$$

satisfies (2) with $c_\gamma = 1$. One checks that it is holomorphic, thus $\rho_f = 0$ and therefore $\partial f/\partial\lambda|_{\lambda=0} = 0$. From this point it is easy to calculate that $\phi = 0$, for example by writing out

$$\phi = \frac{\partial}{\partial\lambda}(\phi_0 + \lambda\phi) = \frac{\partial}{\partial\lambda}[f]$$

in terms of the λ-derivatives of f, f', f'', f'''. The conclusion is that Φ is nonsingular.

For punctured surfaces (with no nontrivial boundary components) the only additional step is the verification that when $\phi_0, \phi \in B_2(G)$ (for G a Fuchsian group which uniformizes S), ρ_f is holomorphic at the punctures of S as well [3]).

For surfaces S with a single hole (and any number of punctures) we have the following result. It may be generalized to arbitrary $m > 0$ in the same manner as Theorem IA (see [4]).

THEOREM 1B. With the same hypothesis as in Theorem 1A, the real analytic mapping $\Phi^* : B_2^*(G) \to \mathrm{Hom}$ *is nonsingular.*

Proof. Define ρ_g as in the proof for the compact case. Note that we allow λ to take on only real values so that $\phi_0 + \lambda\phi$ remains in $B_2^*(G)$. Recall that g_ϕ is normalized at a point of some maximal interval I_0 of $\Omega(G) \cap \mathbb{R}$ and it follows that g_ϕ is real on I_0. We have $\rho_g \in F_{-1}(G)$ and hence by (2), ρ_g is real on $\gamma(I_0)$ for all $\gamma \in G$. Since $m = 1$, ρ_g is real on $\Omega(G) \cap \mathbb{R}$ and extends by reflection to a holomorphic function on $\Omega(G)$. This defines a holomorphic (-1)-differential on $\Omega(G)/G$; thus $\rho_g = 0$ and the proof proceeds as before.

4. *A look at unbounded differentials*

As we remarked earlier, $F_2(G)$ is infinite dimensional when $m + n > 0$. Since Hom is finite dimensional, Φ and Φ' cannot be injective nor nonsingular. Thus one is led to ask the following:

QUESTION. If $S = U/G$ is an open Riemann surface of finite type, is $\Phi : F_2(G) \to \mathrm{Hom}$ surjective?

At any rate, the following is true.

THEOREM 2. Let S *be an open topological surface of finite type and let* $\chi : \pi_1(S) \to \mathrm{Aut}\,\hat{\mathbb{C}}$ *be an arbitrary homomorphism. Then there is a projective structure on* S *whose monodromy homomorphism is precisely* χ.

This can be seen by choosing a Fuchsian group G representing (an arbitrary conformal structure on) S and then constructing by topological methods a local homeomorphism $F: U \to \hat{\mathbb{C}}$ such that (1) holds. Thus f induces a projective structure on S (whose underlying conformal structure need not be the same as that of U/G) with the desired monodromy. The details will appear in [11]. The difficulty of this method is that one is not able to predict how many of the $m+n$ ideal boundary components of S will be punctures and how many will be nondegenerate curves.

For a fixed $\phi_0 \in F_2(G)$, where G is a Fuchsian group of finite type, define $K_{\phi_0} = \text{Ker}\, D_{\phi_0}(\Phi)$. K_{ϕ_0} has finite codimension in $F_2(G)$ and hence is also infinite dimensional. Recalling that $\rho_f \in F_{-1}(G)$ whenever ϕ satisfies (4), we define

$$\Phi_{\phi_0} : K_{\phi_0} \to F_{-1}(G)$$

by the formula

$$\Psi_{\phi_0}(\phi) = \rho_f$$

where $f = f(z,\lambda)$ is a normalized solution of $[f] = \phi_0 + \lambda\phi$. One easily sees that Ψ_{ϕ_0} is linear, and the arguments outlined $(\partial f/\partial\lambda = 0 \Rightarrow \phi = 0)$ show that it is injective. (In fact, the injectivity of Ψ_{ϕ_0} is a direct generalization of the non-singularity of Φ for compact surfaces.) We may go one step further. Let $K'_{\phi_0} = \text{Ker}\, D_{\phi_0}(\Phi')$.

THEOREM 3. *Let* S *be a Riemann surface of finite type and let* G *be a Fuchsian group representing* S. *Let* $\phi_0 \in F_2(G)$ *and suppose that the monodromy group* $\chi_{\phi_0}(G)$ *is not commutative. Then there is an injective linear mapping*

$$\Psi'_{\phi_0} : K'_{\phi_0} \longrightarrow F_{-1}(G)$$

defined by $\Psi'_{\phi_0}(\phi) = \rho_{g_\lambda}$, *where* $(g_\lambda, \eta_\lambda)$ *is any one-parameter family of deformations of* G *such that*

(7) $$[g_\lambda] = \phi_0 + \lambda\phi$$

and

(8) $$(\partial\eta/\partial\lambda)|_{\lambda=0} = 0.$$

Proof. Everything has already been verified except the affirmation that Ψ'_{ϕ_0} is well defined. To see this, let $(g_\lambda, \eta_\lambda)$ and $(\tilde{g}_\lambda, \tilde{\eta}_\lambda)$ be two families of deformations satisfying (7) and (8). Thus

(9) $$\tilde{g}_\lambda = A_\lambda \circ g_\lambda$$

where $A_\lambda(z)$ is a Möbius transformation depending holomorhphically on λ. By a calculation similar to (5a) and (5b) one finds that

(10) $$\rho_{\tilde{g}} = \rho_g + (\rho_A \circ g)(g')^{-1}$$

where here and henceforth we take $\lambda = 0$, and write g, η in place of g_0, η_0.

Since $\rho_g, \rho_{\tilde{g}} \in F_{-1}(G)$ we see that $(\rho_A \circ g)(g')^{-1} \in F_{-1}(G)$ as well. It is easily seen that there are constants $B_0, B_1, B_2 \in \mathbb{C}$ such that

(11) $$\rho_A(z) = B_2 z^2 + B_1 z + B_0, \quad z \in U.$$

For each $\gamma \in G$ there are $a, b, c, d \in \mathbb{C}$, $ad - bc = 1$, such that

(12) $$g \circ \gamma = \eta(\gamma) \circ g = (ag + b)(cg + d)^{-1}.$$

When we substitute (11) and (12) into (10) we find that

$$((a^2-1)B_2 + acB_1 + c^2B_0)g(z)^2 + (2abB_2 + (ad+bc-1)B_1 + 2cdB_0)g(z)$$
$$+ (b^2B_2 + bdB_1 + (d^2-1)B_0) = 0$$

identically for all $z \in U$. Therefore (B_2, B_1, B_0) is an eigenvector of the matrix

$$\begin{pmatrix} a^2 & ac & c^2 \\ 2ab & ad+bc & 2cd \\ b^2 & bd & d^2 \end{pmatrix}$$

with eigenvalue equal to one, and by direct calculation we find that

(13) $$(B_2, B_1, B_0) = (-\mu c, \mu(a-d), \mu b)$$

for some $\mu \in \mathbb{C}$. Let us suppose that $\mu \neq 0$, and consider any $\gamma \in G$ such that $\eta(\gamma)$ is not the identity. If $c \neq 0$ in the notation of (12), then the fixed points of $\eta(\gamma)$ are given by

$$\frac{a-d \pm \sqrt{(a+d)^2-4}}{2c} = \frac{B_1 \pm \sqrt{B_1^2 - 4B_0B_2}}{-2B_2} .$$

If $c = 0$ then the fixed points are either ∞ and

$$\frac{b}{a-d} = \frac{B_1}{B_0}$$

(in case $a \neq d$), or else just ∞ (in case $a = d$). As B_0, B_1, B_2 do not depend on γ we have proved that there are only three possible fixed point sets for $\eta(\gamma)$. The reader will easily verify that this cannot occur in a non-commutative subgroup of Aut $\hat{\mathbb{C}}$; therefore $\mu = 0$. This implies that $\rho_A = 0$, and referring to (10) we find that $\rho_{\tilde{g}} = \rho_g$, so Ψ' is indeed well defined.

We close with some questions, the answers to which could shed some light on the Schwarzian differential equation.

1. Does K'_{ϕ_0} properly contain K_{ϕ_0}? (This relates to the dependence of Φ on the choice of base point in the normalization of f.)

2. Is Ψ'_{ϕ_0} surjective? (An explicit construction of a $\phi \in K'_{\phi_0}$ which is sent to a given $\rho \in F_{-1}$ would be of particular interest.)

3. How does K_{ϕ_0} vary with ϕ_0?

CENTRO DE INVESTIGACION Y DE ESTUDIOS AVANZADOS
MEXICO CITY

REFERENCES

[1] P. Appel, E. Goursat and P. Fatou, *Théorie des Functions Algébriques*, Vol. 2, Gauthier-Villars, 1930.

[2] L. Bers, "Nielsen extensions of Riemann surfaces," *Ann. Acad. Sci. Fenn.* AI 2 (1976), 29-34.

[3] D. Gallo and R. Porter, "Embedding the deformation space of a Fuchsian group of the first kind," to appear, *Bol. Soc. Mat. Mex.* 26 no. 2 (1981).

[4] D. Gallo and R. Porter, "Projective structures on bordered surfaces," to appear.

[5] R.C. Gunning, "Analytic structures on the space of flat vector bundles over a compact Riemann surface," *Several Complex Variables II*, Maryland, 1979. Springer Lecture Notes 185 (1971), 47-62.

[6] R.C. Gunning, "Special coordinate coverings of Riemann surfaces," *Math. Ann.* 170 (1967), 67-86.

[7] I. Kra, "On affine and projective structures on Riemann surfaces," *J. d'Analyse Math.* 22 (1969), 285-298.

[8] I. Kra, "Deformations of Fuchsian groups II," *Duke, Math. J.* 38 (1971), 499-508.

[9] I. Kra, "A generalization of a theorem of Poincaré," *Proc. Amer. Math. Soc.*, 27 (1971), 299-302.

[10] H. Poincaré, "Mémoire sur les fonctiones fuchsiennes," *Acta Math.* 1 (1882), 193-294.

[11] R.M. Porter, "Projective surfaces with prescribed monodromy," this volume.

THE TEICHMULLER-KOBAYASHI METRIC FOR INFINITE DIMENSIONAL COMPLEX TEICHMULLER SPACES

FREDERICK P. GARDINER*

Introduction

The purpose of this paper is to give an expository account of some of the basic theorems of Teichmüller theory. We include Royden's theorem on the equality of Teichmüller's and Kobayashi's metrics. We also prove the new result that this theorem extends to infinite dimensional Teichmüller spaces. Also, we give a direct proof, following from the main inequality of Reich and Strebel [13,15], that Teichmüller's metric is the integral of its differentiated form. The proof we give works for infinite dimensional Teichmüller spaces modelled on Fuchsian groups. Universal Teichmüller space is included as the special case when the Fuchsian group consists of the identity alone.

An important open problem is to determine whether in the infinite dimensional case a biholomorphic self-mapping h of Teichmüller space T is induced by an element of the modular group. The equality of Kobayashi's and Teichmüller's metrics shows that h is an isometry in Teichmüller's metric and that the derivative of h yields a bijective isometry on each fibre of the tangent bundle of T. The problem would be solved if one could show that each such bijective isometry is induced by a conformal mapping of the base Riemann surfaces. For the finite dimensional cases Royden proves this in [14] by analyzing the smoothness of the expression $\iint|\phi_0 + t\phi_1|dxdy$ as a function of t for certain quadratic differentials ϕ_0 and ϕ_1. This method does not seem to adapt in any obvious way to the infinite dimensional cases.

In writing the section on Royden's theorem for the finite dimensional case, I have benefited from lecture notes of C.J. Earle. In particular, Earle's modified version of Ahlfors' lemma, Lemma 3.6, was helpful.

*This research was supported in part by a grant from the City University of New York PSC-CUNY Research Award Program.

1. *Preliminaries*

Let Γ be a non-elementary Fuchsian group with limit set $\Lambda \subseteq \hat{\mathbb{R}}$ where $\hat{\mathbb{R}} = \mathbb{R} \cup \{\infty\}$ and let C be a closed set invariant under Γ with $\Lambda \subseteq C \subseteq \hat{\mathbb{R}}$. Let $M(\Gamma)$ be the set of all complex-valued L_∞ functions μ with support in the upper half-plane U such that

(1.1) i) $\|\mu\|_\infty < 1$ and

ii) $\mu(Az)\overline{A'(z)} = \mu(z)A'(z)$ for all A in Γ.

$M(\Gamma)$ is called the space of Beltrami coefficients. To each μ in $M(\Gamma)$ let $\tilde{\mu}$ be the extension of μ to the lower half-plane given by the rule $\tilde{\mu}(\bar{z}) = \overline{\mu(z)}$. Let w_μ be the unique quasiconformal homeomorphism of the extended complex plane [3] which satisfies

(1.2) iii) $\frac{\partial}{\partial \bar{z}} w_\mu = \tilde{\mu} \frac{\partial}{\partial z} w_\mu$ in the sense of distributions and

iv) $w_\mu(0) = 0$, $w_\mu(1) = 1$, $w_\mu(\infty) = \infty$.

The Teichmüller space $T = T(\Gamma)$ is defined to be a set of equivalence classes of elements of $M(\Gamma)$. Two elements μ and υ are equivalent $(\mu \sim \upsilon)$ if $w_\mu(x) = w_\upsilon(x)$ for all x in C. T depends on Γ and C but we suppress this dependency in the notation.

Let $\Omega = \hat{\mathbb{C}} - C$ and $A = A_s(\Omega,\Gamma)$ be space of holomorphic functions ϕ in Ω satisfying

(1.3) i) ϕ is symmetric, that is, $\phi(\bar{z}) = \overline{\phi(z)}$,

ii) $\phi(Az)A'(z)^2 = \phi(z)$ for all A in Γ,

iii) $\|\phi\| = \iint_\omega |\phi| \, dxdy < \infty$.

In this last condition ω is a fundamental domain for Γ acting on U.

Given μ in $M(\Gamma)$ let $k_0 = \inf\{\|\nu\|_\infty ; \nu \sim \mu\}$ and $K_0 = (1+k_0)/(1-k_0)$. The following inequality is proved in [13,15,16,4].

$$\frac{1}{K_0} \leq \iint_\omega |\phi| \frac{|1+\mu\phi/|\phi||^2}{1-|\mu|^2} dxdy \tag{1.4}$$

for any ϕ in A with $\|\phi\| = 1$. There is an equally important companion inequality:

$$K_0 \leq \sup \iint_\omega |\phi| \frac{|1-\mu\phi/|\phi||^2}{1-|\mu|^2} dxdy \tag{1.5}$$

where the supremum is over all ϕ in A with $\|\phi\| = 1$. The inequality (1.5) is proved in [13,16] and for infinitely generated groups Γ in [9].

Following the notation of Reich and Strebel in [13], we introduce the functionals

$$H[\mu] = \sup \operatorname{Re} \iint_\omega \phi\mu dxdy,$$

$$I[\mu] = \sup \operatorname{Re} \iint_\omega \phi \frac{\mu}{1-|\mu|^2} dxdy, \text{ and} \tag{1.6}$$

$$J[\mu] = \sup \iint_\omega |\phi| \frac{|\mu|^2}{1-|\mu|^2} dxdy$$

where these suprema are taken over all ϕ in A with $\|\phi\| = 1$. From (1.4) one can easily show that if $I[\mu] = k_0/(1-k_0^2)$ then μ is extremal. From (1.5) one can prove the converse; if μ is extremal then $I[\mu] = k_0/(1-k_0^2)$.

Similarly from (1.4) and (1.5) one finds that μ is extremal if, and only if, $H[\mu] = k_0$. This is known as Hamilton's codition, [10]. All of these proofs are given in [13]. Also in [13] Reich and Strebel present the following important inequality,

$$-\left\{J[\mu] - \frac{k_0^2}{1-k_0^2}\right\} + \frac{k_0}{1-k_0^2} \leq I[\mu] \leq \frac{k_0}{1-k_0^2} + \left\{J[\mu] - \frac{k_0^2}{1-k_0^2}\right\}. \tag{1.7}$$

The right-hand side of (1.7) is a direct consequence of (1.4) and the left-hand side is a consequence of (1.5).

2. *Teichmüller's metric*

In this section the objective is to show how the inequality (1.7) of Reich and Strebel can be used to show that Teichmüller's metric is the integral of its differentiated form. The result is due to O'Byrne [12]. His proof depends on a general lifting theorem for quotients of Finsler manifolds and on the paper of Earle and Eells [7] which shows that Teichmüller's metric is a quotient Finsler metric with respect to the Bers foliation for Teichmüller space. The proof presented here follows a different course and is more direct.

Let μ be in $M(\Gamma)$ and μ_0 be extremal in the class of μ and $k_0 = \|\mu_0\|_\infty$. By a normal families argument for quasiconformal mappings of bounded dilation such an extremal element always exists. By definition the Teichmüller distance from $[0]$ to $[\mu]$ is given by

$$d([0],[\mu]) = \frac{1}{2}\log((1+k_0)/(1-k_0)). \tag{2.1}$$

To extend this definition so that it gives the distance between two arbitrary points, one can use the translation mapping induced by composition. Explicitly, let $\mu \in M(\Gamma)$ and $\Gamma_\mu = w_\mu \Gamma w_\mu^{-1}$. Γ_μ will also be a Fuchsian group acting on the upper half-plane. To show that the A_μ determined by

$$A_\mu \circ w_\mu = w_\mu \circ A \tag{2.2}$$

is a Möbius transformation, one first observes that it is a homeomorphic self-mapping of U. Then one shows from (1.1) that both w_μ and $w_\mu \circ A$ have the same Beltrami coefficient. Thus for a certain linear fractional mapping A, both sides of (2.2) will have the same normalization and the same Beltrami coefficient. By the uniqueness of solutions to (1.2) with the same normalization, this shows (2.2) holds with A_μ of the form $(az+b)/(cz+d)$ where $ad-bc = 1$.

The right translation mapping

$$R_\mu^{-1} : M(\Gamma) \longrightarrow M(\Gamma_\mu)$$

is defined by letting $(R_\mu^{-1})(\sigma)$ be the Beltrami coefficient of $w_\sigma \circ (w_\mu)^{-1}$. If we let $C_\mu = w_\mu(C)$ and C_μ determine the equivalence relation on $M(\Gamma_\mu)$ in a way analogous to the way C determines the equivalence relation on $M(\Gamma)$, then R_μ^{-1} preserves equivalence classes and induces a mapping from $T(\Gamma)$ to $T(\Gamma_\mu)$. We use the same symbol R_μ^{-1} to denote the induced mapping. Notice that $R_\mu^{-1}[\mu] = 0$. Now, by definition,

$$d([\mu],[\nu]) = d([0],R_\mu^{-1}[\nu]). \tag{2.3}$$

It is clear that the translation mapping R_μ^{-1} is an isometry under the Teichmüller metrics on $T(\Gamma)$ and $T(\Gamma_\mu)$.

To find the infinitesimal form of Teichmüller's metric, one first calculates $d([0],[t\mu])$ to first order in the real variable t. Let $k_0(t)$ be the extremal value of the sup norm of a Beltrami coefficient in the equivalence class of t. From (2.1) $k_0(t)$ has the same derivative at $t = 0$ as $d([0],[t\mu])$. Now the inequalities of (1.7) enable us to evaluate $k_0'(0)$. The curly bracket terms in (1.7) are of order t^2, as is obvious from inspection of the integrand in the definition of $J[\mu]$. Moreover, if the denominator in the integrand for $I[\mu]$ is replaced by 1, an error only of order t^2 will be introduced. These considerations yield, for $t > 0$,

$$d(0,[t\mu]) = t \sup \operatorname{Re} \iint \mu\,\phi\, dxdy + O(t^2) \tag{2.4}$$

where the supremum is over all ϕ in A with $\|\phi\| = 1$. We remark that this proof of (2.4) depends on knowing inequality (1.5) as well as (1.4).

Formula (2.4) can be generalized to give the infinitesimal length $F([\mu],\nu)$ of a tangent vector ν at an arbitrary point $[\mu]$. By definition, $F([\mu],\nu)$ is the derivative with respect to t at $t = 0$ of the function $d([\mu],[\mu+t\nu])$. Replacing ν by $\mu + t\nu$ in (2.3) we get

$$d([\mu],[\mu+t\nu]) = d(0,R_\mu^{-1}(\mu+t\nu)) = \sup \operatorname{Re} \iint t\phi S(\nu)dudv + O(t^2)$$

where S is the derivative at μ of R_μ^{-1} and $u + iv = w = w_\mu$. Here the integral is over a fundamental domain ω_μ for $\Gamma = w_\mu \Gamma w_\mu^{-1}$ and the supremum is over all ϕ in A with $\|\phi\| = 1$. Now

$$R_\mu^{-1}(\sigma) = \left[\frac{\sigma - \mu}{1 - \bar{\mu}\sigma}\ \frac{1}{\theta}\right] \circ w_\mu^{-1} \tag{2.5}$$

where $\theta = \bar{p}/p$ and $p = \frac{\partial}{\partial z} w_\mu$. Letting $\sigma = \mu + t$, we find $S(\upsilon) = \upsilon/(1-|\mu|^2)\theta$ and

$$(2.6) \qquad F([\mu],\upsilon) = \sup \operatorname{Re} \iint_{\omega_\mu} \phi(w) \left[\frac{\upsilon}{1-|\mu|^2} \cdot \frac{1}{\theta} \right] dudv$$

where the supremum is over all ϕ in $A_\mu = A_s(\Omega_\mu, \Gamma_\mu)$ for which $\|\phi\| = 1$.

In Lemma 3.1 we show that F is C^1 on the tangent bundle to $T(\Gamma)$ for finite dimensional Teichmüller spaces. The following lemma applies to infinite dimensional Teichmüller spaces as well.

LEMMA 2.1. The function F *from the tangent bundle of* $T(\Gamma)$ *to* $\mathbb{R}$ *is continuous.*

Proof. Since right translation R_μ is continuous, it suffices to show F is continuous at $\mu = 0$. In other words, we must show $F([0],\nu)$ is close to $F([\mu],\nu_1)$ when $\|\mu\|_\infty < \varepsilon$ and $\|\upsilon - \upsilon_1\|_\infty < \varepsilon$. Because F is a semi norm in υ it is enough to show $F([0],\nu)$ is close to $F([\mu],\nu)$ for $\|\mu\|_\infty < \varepsilon$ and $\|\upsilon\|_\infty < 1$. From (2.6) we see that given $\varepsilon > 0$ and ϕ in A with $\iint_{\omega_\mu} |\phi| = 1$, we must find ψ in A with $\iint_\omega |\psi| = 1$ such that

$$(2.7) \qquad \operatorname{Re} \iint_{\omega_\mu} \phi(w) \left| \frac{\upsilon}{1-|\mu|^2} \cdot \frac{1}{\theta} \right| dudv < \operatorname{Re} \iint_\omega \psi \upsilon dxdy + \varepsilon.$$

Proving (2.7) will be sufficient because the same method of proof will yield an opposite inequality. Let Θ_μ be the theta series operator for the group Γ_μ. That is, $\Theta_\mu F = \Sigma F(Az)A'(z)^2$ where the summation is over all A in Γ_μ. Since Θ_μ is surjective [5] there is an integrable holomorphic function F in Ω, real on the real axis, for which $\phi = \Theta_\mu F$ and the left side of (2.7) can be rewritten as

$$(2.8) \qquad \operatorname{Re} \iint_U F \left| \frac{\upsilon}{1-|\mu|^2} \cdot \frac{1}{\theta} \right| dudv.$$

By the Lebesgue dominated convergence theorem, (2.8) can be made arbitrarily close to $\operatorname{Re} \iint_U \upsilon F dxdy$ if $\|\mu\|_\infty$ is sufficiently small. Moreover,

$$\iint_U \upsilon F dxdy = \iint_\omega \upsilon \Theta F dxdy$$

when $\Theta = \Theta_0$. Our candidate for ψ in (2.7) is $c_\mu \Theta F$ where c_μ is a positive constant. We have to show $c_\mu \longrightarrow 1$. It is enough to prove the following lemma.

LEMMA 2.2. $\|\Theta_\mu F\| \longrightarrow \|\Theta F\|$ *as* $\|\mu\|_\infty \longrightarrow 0$.

Proof. Let B_μ^n be an enumeration of Γ_μ and $\Theta_{\mu n}$ be the truncation of Θ_μ to the first n elements of Γ_μ. Clearly $\iint_{\omega_\mu} |\Theta_{\mu n} F| \longrightarrow \iint_\omega |\Theta_n F|$ as $\|\mu\|_\infty \longrightarrow 0$. In order to pass to the limit, we show that for any $\varepsilon > 0$, there exists n_0 and $\delta > 0$, such that for $n \geq n_0$ and $\|\mu\|_\infty < \delta$

$$\iint_{\omega_\mu} \sum_{n > n_0} |F(B_\mu^n z) B_\mu^n{}'(z)^2| < \varepsilon. \tag{2.9}$$

To simplify notation let U be replaced by the unit disk Δ and assume the groups Γ and Γ_μ act on Δ. Pick $r < 1$ so that $\iint_{r<|z|<1} |F| < \varepsilon$ and pick n_0 so that $D = \bigcup_{n=1}^{n_0} B^n(\omega)$ contains the disk of radius $(r+1)/2$. Now $\omega_\mu = w_\mu(\omega)$ and hence $w_\mu(D) = \bigcup_{n=1}^{n_0} B_\mu^n(\omega_\mu)$. Since w_μ satisfies a Hölder condition [1] of order $\alpha = \frac{1}{K}$ where $K = (1 + \|\mu\|_\infty)/(1 - \|\mu\|_\infty)$, we see that for sufficiently small $\|\mu\|_\infty$, $w_\mu(D) \supseteq \{z; |z| < r\}$. This implies that (2.9) is bounded by $\iint_{r<|z|<1} |F|$ which is less than ε.

This completes the proof of Lemmas 2.1 and 2.2.

Now let $\bar{d}$ be the integrated form of (2.6). This means $\bar{d}(p,q) = \inf L(\gamma)$ where γ is a piecewise smooth path joining p to q, that is $\gamma(0) = p$ and $\gamma(t_0) = q$, and $L(\gamma) = \int_0^{t_0} F(\gamma(t), \gamma'(t))dt$. It is a general fact and easy to prove that if F is a continuous function on the tangent bundle, then $d \leq \bar{d}$.

THEOREM 2.1. For any $T(\Gamma)$, $d = \bar{d}$.

This theorem is elementary for finite dimensional Teichmüller spaces and is mentioned in [14, page 370]. The general case is treated by O'Byrne [12, page 326] by means of a general theorem concerning quotients of Finsler structures.

We must show that $\bar{d}(0,[\mu]) \leq d(0,[\mu])$. Assume $\|\mu\|_\infty = 1$ and $k\mu$ is extremal. By definition $d(0,[k\mu]) = \frac{1}{2}\log(1+k)/(1-k)$. Let $\gamma(t) = [t\mu]$, $0 \leq t \leq k$. We will show $L(\gamma) = d(0,[k\mu])$. Since $\gamma'(t) = \mu$, we must calculate $F([t\mu],\mu)$. We know that $w_{t\mu}$ and $(w_{t\mu})^{-1}$ are both extremal. The Beltrami coefficient of $(w_{t\mu})^{-1}$ is $-t\mu\frac{1}{\theta}$ where $\theta = \bar{p}/p$ and $p = \frac{\partial}{\partial z} w_{t\mu}$. Since $-t\mu\frac{1}{\theta}$ is extremal, Hamilton's condition tells us that

$$\sup \operatorname{Re} \iint_{\omega_{t\mu}} \phi \frac{t\mu}{1-t^2|\mu|^2} \cdot \frac{1}{\theta}\, du\, dv = \frac{t}{1-t^2} \tag{2.10}$$

where the supremum is over ϕ in A_μ with $\|\phi\| = 1$. Substituting (2.10) into (2.6) we get

$$F([t\mu],\mu) = \frac{1}{1-t^2}$$

and, therefore,

$$\int_0^k F(\gamma(t),\gamma'(t)) = \int_0^k \frac{dt}{1-t^2} = \frac{1}{2}\log\frac{1+k}{1-k}. \qquad \text{Q.E.D.}$$

3. *Royden's theorem in the finite dimensional case*

In this section we assume $C = \hat{\mathbf{R}}$ so that T has a complex structure. The upper half plane U and the lower half plane L are each components of Ω. In this case, restriction to U gives a real linear isomorphism of the space $A_s(\Gamma,\Omega)$ onto the space $A(\Gamma,U)$, the complex linear space of integrable holomorphic quadratic differentials with support in U. The space $A(\Gamma,U)$ is the cotangent space to T at the origin and part iii) of (1.3) gives a cometric on this cotangent space, which by (2.4) is the dual metric to the infinitesimal form of Teichmüller's metric. As in Sections 1 and 2, we adopt the shortened notation $A = A_s(\Gamma,\Omega) = A(\Gamma,U)$.

The proof of the next lemma is of a technical nature. If the reader is willing to accept it without proof, he should skip to the first paragraph after formula (3.5).

LEMMA 3.1. *Suppose* A *is finite dimensional. Then the metric* F *given by* (2.6) *is* C^1 *on the tangent bundle to* T *except at the zero section.*

Proof. We outline the proof given by Royden in [14, page 374]. The first observation is that F is dual to the cometric G on the cotangent space to T given by the formula

$$(3.1)\qquad G([\mu],\phi) = \iint_{\omega_\mu} \left| \frac{\phi(w)}{(1-|\mu|^2)} \right| dudv = \iint_{\omega} |\phi(w(z))w_z^2(z)|dxdy$$

for μ in $M(\Gamma)$, ϕ in A_μ and $w = u+iv = w_\mu$. Since the dual of a strictly convex C^1 metric is a C^1 metric [6], the lemma will be proved if we show (3.1) is C^1 except at the zero section. (In [14] Royden proves the more difficult fact that (3.1) is $C^{1+2/n}$ where n is the largest possible order of a zero of an element of A. This follows from his main lemma of [14, Lemma 1].) To see that (3.1) is C^1, first consider the case of calculating the derivative of $G([\mu],\phi_0 + t\phi_1)$, (where ϕ_0 is not zero) with respect to t at $t = 0$. It is straightforward to show that it is

$$(3.2)\qquad \mathrm{Re} \iint_{\omega_\mu} \frac{|\phi_0|}{\phi_0}\, \phi_1 \frac{1}{1-|\mu|^2}\, dudv.$$

It is clear that this derivative depends continuously on ϕ_0 in $A(\Gamma_\mu,U)$ as long as $\phi_0 \neq 0$. In order to see how (3.1) depends on $[\mu]$ we must introduce a local coordinate for the cotangent space to T. Since T is a complex analytic manifold modelled on A we can find an open set U_1 in T containing μ_0 and an open set U_2 in A such that the cotangent bundle to U_1 is isomorphic to $U_2 \times A$. In articular, there is a linear mapping $L_\mu : A_{\mu_0} \longrightarrow A$ depending holomorphically on μ uch that

$$U_2 \times A \ni ([\mu],\psi) \longmapsto ([\mu], L_\mu(\psi)) \in T^*T \tag{3.3}$$

s a chart for the cotangent bundle. Substituting (3.3) into (3.1) and letting $_\mu = w_{\mu_1} \circ w_{\mu_0}$ and $w_1 = w_{\mu_1}$ and $w_0 = w_{\mu_0}$, we get

$$G([\mu],\psi) = \iint_{\omega_{\mu_0}} \frac{|L_\mu(\psi)(w_1)||p_1|^2 |1+\mu_1\theta\mu_0|^2}{1-|\mu_0|^2}\, du_0 dv_0 \tag{3.4}$$

here $p_1 = \dfrac{\partial w_1}{\partial w}$, $p_0 = \dfrac{\partial w_0}{\partial z}$ and $\theta = \bar{p}_0/p_0$. Since L_μ is holomorphic in μ, it can e written as a power series in the variable μ_1:

$$L_\mu(\psi) = L_{\mu_0}(\psi) + \mu_1\eta + O(\mu_1^2).$$

ere, η is an L_1-quadratic differential and η depends continuously on μ_0. Letting (t) be defined by $w_{\mu(t)} = w_{t\mu_1} \circ w_{\mu_0}$ we get from (3.4)

$$G([\mu(t)],\psi) = \iint_{\omega_{\mu_0}} \frac{|L_{\mu_0}(\psi) + t\mu_1\eta_1|}{1-|\mu_0|^2}\, du_0 dv_0 \tag{3.5}$$

here η_1 is an L_1-quadratic differential depending continuously on μ_0 (but η_1 s not necessarily holomorphic). From (3.2) and (3.5) it follows that G is C^1 in oth coordinates as long as the second coordinate is non-zero and the Teichmüller pace is finite dimensional.

Since T can be realized as an open domain in $\mathbb{C}^n$ we will now follow Royden nd use coordinate notation. An element of the tangent bundle will be a pair $(x;\xi)$ here x is in the base space and ξ is a tangent vector. We also will write $= (x_1, x_i)$ where $2 \leq i \leq n$ and, similarly, $\xi = (\xi_1, \xi_j)$, $2 \leq j \leq n$. Always, i nd j will be integers ranging from 2 to n. We will write the Taylor series of F up to first order and then adjust the coordinates $(x;\xi)$ so that F has a special form. Our first step is to pick the first coordinate x_1 so that

$$F(x_1,0;1,0) = \frac{1}{1-|x_1|^2}\,. \tag{3.6}$$

This is achieved by letting $(x_1,0)$ correspond to a Beltrami differential of the form $x_1 \frac{|\phi|}{\phi}$ where ϕ is any holomorphic integrable quadratic differential. We have freedom to let the other coordinates of the chart x take any values so long as x determines a local homeomorphism of a coordinate patch.

The Taylor expansion for $F(x_1, x_i; 1, \xi_j)$ is

$$F(x_1, 0;1,\xi_j) + \mathrm{Re}\,\Sigma\, C_i(x_1, \xi_j)x_i + O(|x_i|). \tag{3.7}$$

Expanding $F(x_1, 0;1,\xi_j)$ relative to the variable ξ_j, we get

$$F(x_1, 0;1,\xi_j) = F(x_1, 0;1,0) + \mathrm{Re}\,\Sigma\, B_j(x_1)\xi_j + O(|\xi_j|). \tag{3.7'}$$

Here, B_j and C_i are continuous functions and the error estimates $O(|x_i|)$ and $O(|\xi_j|)$ hold uniformly in a neighborhood of the origin in the variables x_1 and ξ_j for formula (3.7) and in the variable x_1 for formula (3.7'). Now, make the change of coordinates

$$\begin{aligned} x_1 &= y_1 - \mathrm{Re}\,\Sigma\, B_i(0)y_i \\ x_i &= y_i\,, \end{aligned} \tag{3.8}$$

which on the fibers of the tangent bundle yields

$$\begin{aligned} \xi_1 &= \eta_1 - \mathrm{Re}\,\Sigma\, B_j(0)\eta_j \\ \xi_j &= \eta_j\,. \end{aligned} \tag{3.9}$$

Notice that (3.6) is preserved for the (y,η) coordinates and $\frac{\partial F}{\partial \eta_j}(x_1,0;\xi_1,\xi_j) =$ $\frac{\partial F}{\partial \xi_1}(-\operatorname{Re} B_j(0)) + \frac{\partial F}{\partial \xi_j}$. But $\frac{\partial F}{\partial \xi_j}(x_1,0;1,0) = \operatorname{Re} B_j(x_1)$ and $\frac{\partial F}{\partial \xi_1}(x_1,0;\xi_1,0) =$ $\frac{1}{1-|x_1|^2}$. Hence $\frac{\partial F}{\partial \eta_j}(0,0;1,0) = 0$. This means that for the coordinates (y,η) the functions $B_j(y_1)$ satisfy $B_j(0) = 0$.

So we can assume (3.6) holds for the coordinates (x,ξ) with $B_j(0) = 0$. We now make a second change of coordinates which does not alter these properties and for which $C_i(0,0) = 0$. Indeed, let

(3.10)
$$x_1 = y_1 - \operatorname{Re} \Sigma\, C_i(0,0) y_i y_1 ,$$
$$x_i = y_i ,$$

which on the fibers of the tangent bundle yields

(3.11)
$$\xi_1 = \eta_1 - \operatorname{Re} \Sigma\, C_j(0,0) y_j \eta_1 - \operatorname{Re} \Sigma\, C_j(0,0) \eta_j y_1$$
$$\xi_j = \eta_j .$$

Again, notice that when $x_i = 0$ and $\xi_j = 0$ we have $x_1 = y_1$ and $\xi_1 = \eta_1$ so (3.6) goes over to the new coordinates. Moreover, in the $(y;\eta)$ coordinates we have

(3.12)
$$\frac{\partial F}{\partial \eta_j}(y_1,0;1,\eta_j) = \frac{\partial F}{\partial \xi_1} \operatorname{Re}(-C_j(0,0)y_1) + \frac{\partial F}{\partial \xi_j} .$$

When we evaluate (3.12) at $(y,\eta) = (0,0;1,0)$ we see that because in the first change of coordinates we achieved $\frac{\partial F}{\partial \xi_j}(0,0;1,0) = 0$, we have $\frac{\partial F}{\partial \eta_j}(0,0;1,0) = 0$. Furthermore,

(3.13)
$$\frac{\partial F}{\partial y_i} = \frac{\partial F}{\partial x_i} \operatorname{Re}(-C_i(0,\,0)y_1) + \frac{\partial F}{\partial x_i} + \frac{\partial F}{\partial \xi_1} \operatorname{Re}(-C_i(0,0)\eta_1).$$

Evaluating (3.13) at $(y,\eta) = (0,0;1,0)$, the first term on the right is zero, the second is $+C_i(0,0)$ and the third is $-C_i(0,0)$. We have proved the following lemma.

LEMMA 3.2. Given any unit tangent vector $|\phi|/\phi$ *to* T *at the origin, there is a choice of coordinates* $(x,\xi) = (x_1,x_i;\xi_1,\xi_i)$ *such that*

a) $(x_1,0) = [x_1|\phi|/\phi]$ *for* $x_1 \in \Delta$,

b) $F(x_1,0;1,0) = \dfrac{1}{1-|x_1|^2}$ *for* $x_1 \in \Delta$,

c) $B_j(0) = 0$ *for* $2 \leq j \leq n$ *in* (3.7') *and*

d) $C_i(0,0) = 0$ *for* $2 \leq i \leq n$ *in* (3.7).

LEMMA 3.3. For the coordinates of Lemma (3.2), $B_i(x_1) = O(|x_1|)$ *as* $x_1 \to 0$.

Proof. Let $g(t)$ be a smooth function from $[0,1]$ into C^n with $g(0) = g(1) = 0$ and for which $g_1(t)$ is identically zero. Let $\gamma(t) = t\zeta e_1$ where $e_1 = (1,0,0,\ldots,0)$ and $0 \leq t \leq 1$ and ζ is in Δ. By part a) of Lemma 3.2 γ is a geodesic path and the distance from 0 to ζe_1 is less than or equal to the length of the path $\gamma + \varepsilon g$. Thus, so long as ε is small enough so that the path $\gamma + \varepsilon g$ is in the same coordinate neighborhood and $\varepsilon > 0$, one has

$$0 \leq \lim_{\varepsilon \to 0+} \int_0^1 \frac{F(t\zeta e_1 + \varepsilon g(t), \zeta e_1 + \varepsilon g'(t)) - F(t\zeta e_1, \zeta e_1)}{\varepsilon}\, dt.$$

By applying (3.7) and (3.7') and using the fact that F is positive homogeneous in ξ, one finds that

$$0 = \int_0^1 \operatorname{Re}(B_j(t\zeta)g_j'(t) + \zeta C_j(t\zeta,0)g_j(t))\,dt. \tag{3.14}$$

Since g can be multiplied by an arbitrary complex number of absolute value 1, we can drop the real part symbol in (3.14). Thus for all C^1 functions g_j with $g_j(0) = g_j(1) = 0$ we get $\int_0^1 h(t)g_j'(t)\,dt = 0$ where $h(t) = B_j(t\zeta) - \zeta\int_0^t C_j(s\zeta,0)\,ds$.

Thus $h(t)$ is constant. It is equal to 0 because $h(0) = 0$ and so

$$B_j(\zeta) = \zeta \int_0^1 C_j(s\zeta,0)ds. \tag{3.15}$$

But (3.15) is $O(|\zeta|)$ since C_j is continuous.

LEMMA 3.4. $F(x_1,0;1,\xi_i) \geq F(x_1,0;1,0) + \operatorname{Re}\Sigma B_i(x_1)\xi_i$.

Proof. Let $h(t) = F(x_1,0;1,t\xi_i)$. It is a convex C^1 function of t. Thus $h'(0) \leq h(1) - h(0)$ and the lemma follows from (3.7').

REMARK. This inequality is the starting point for constructing a supporting metric in Theorem 3.1. It is a different inequality for different choices of local coordinate. By taking coordinates satisfying Lemma 3.2 we get the right inequality.

LEMMA 3.5. Let f *be a holomorphic function from* Δ *into* T *and suppose* $f(0) = 0$. *Let* $\lambda(\zeta) = F(f(\zeta);f'(\zeta))$. *Then there exists a smooth function* $\lambda_0(\zeta)$ *defined for* ζ *in a neighborhood of* 0 *such that* $\lambda_0(0) = \lambda(0) = 1$, $\lambda_0(\zeta) \leq \lambda(\zeta) + O(\zeta^2)$ *and* $-(\Delta \log \lambda_0)/\lambda_0^2 \leq -4$.

Proof. The expression $-(\Delta \log \lambda)/\lambda^2$ where Δ is the Laplacian is called the curvature of $\lambda(\zeta)|d\zeta|$. The curvature is invariant under holomorphic change of coordinates. Now suppose $f(z)$ from Δ into T is given with $f(0) = 0$ and $f'(0) = k|\phi|/\phi$ where ϕ is a holomorphic quadratic differential. Pick local coordinates for T which satisfy the conditions of Lemma 3.2 for the unit tangent vector $|\phi|/\phi$ and let $f = (f_1,f_2,\ldots,f_n)$ in these coordinates. Clearly, from a) of Lemma 3.2, $f_1'(0) \neq 0$ and $f_i'(0) = 0$. Now alter the local parameter at the origin in Δ so that $f_1(\zeta) = \zeta$ and $f_i(\zeta) = a_i\zeta^2 + O(\zeta^3)$. Then $\lambda(\zeta) = F(\zeta,f_i(\zeta);1,f_i'(\zeta))$, and from Lemma 3.4 and the first part of (3.7) we have

$$\lambda(\zeta) \geq F(\zeta,0;1,0) + \operatorname{Re}\Sigma B_i(\zeta)(2a_i\zeta) + \operatorname{Re}\Sigma C_i(\zeta,f_i'(\zeta))a_i\zeta^2 + O(|\zeta^2|).$$

Using Lemma 3.3, we have

$$\lambda(\zeta) \geq F(\zeta,0;1,0) + O(|\zeta^2|)$$

and since $F(\zeta,0;1,0) = \frac{1}{1-|\zeta|^2}$ which has curvature -4, this proves the lemma.

The following lemma comes from Ahlfors, [2]. We present a modified version coming from unpublished notes of Earle. If $\lambda|dz|$ is a conformal metric on the disk, λ_0 is called a supporting metric for λ at p if $\lambda_0(p) = \lambda(p)$ and if $\lambda_0(\zeta) \leq \lambda(\zeta) + O(|\zeta^2|)$ for a local parameter ζ centered at p.

LEMMA 3.6. *Let* $\lambda|dz|$ *be a conformal metric on the unit disk* Δ *such that at each point* p *in* Δ *there is a smooth supporting metric for* λ *with curvature which is at most* -4. *Then* $\lambda(z) \leq (1-|z|^2)^{-1}$, *that is,* λ *is bounded by the Poincaré metric on the unit disc.*

The proof of this follows exactly as in Ahlfors [2]. Adding a term of order $O(|\zeta|^2)$ in the definition of supporting metric does not invalidate Ahlfors' proof.

THEOREM 3.1. *Let* f *be a holomorphic function from* Δ *into* T. *Then*

$$F(f(\zeta), f'(\zeta)) \leq \frac{1}{1-|\zeta|^2}$$

Proof. Let $\lambda(\zeta) = F(f(\zeta), f'(\zeta))$. Let $p = f(\zeta_0)$. Then let the Teichmüller space T be modelled by the Beltrami differentials on the marked Riemann surface represented by p. For this model the zero Beltrami differential corresponds to the point p itself. A change of coordinate taking ζ_0 to zero will not change the curvature of λ. Hence by Lemma 3.6 there is a smooth supporting metric for λ at ζ_0 with curvature ≤ -4. Hence the theorem follows from Lemma 3.6.

Now let $d_K(p,q)$ be the function defined on Teichmüller space by

$$(3.16) \qquad d_K(p,q) = \inf \rho(f^{-1}(p), f^{-1}(q))$$

where ρ is the Poincaré metric on the unit disk Δ and the infimum is over all holomorphic functions f from Δ into T.

THEOREM 3.2 (Royden). $d_K(p,q) = d(p,q)$, *that is, the infimum of (3.16) is identical to Teichmüller's metric. As a consequence Kobayashi's and Teichmüller's metrics coincide on finite dimensional Teichmüller spaces.*

Proof. Let f be a holomorphic function from Δ into T with $f(\zeta_0) = p$ and $f(\zeta_1) = q$. From Theorem 3.1 we know that $F(f(\zeta), f'(\zeta)) \leq \frac{1}{1-|\zeta|^2}$. Let α be the Poincaré geodesic in Δ joining ζ_0 to ζ_1. From this inequality we see that the Teichmüller length of $f(\alpha)$ is less than or equal to the Poincaré distance from ζ_0 to ζ_1. Thus, from Theorem 2.1, the Teichmüller distance from $f(\zeta_0)$ to $f(\zeta_1)$ is less than the Poincaré distance from ζ_0 to ζ_1 and this shows $d(p,q) \leq d_K(p,q)$.

On the other hand, just as in the proof of Theorem 3.1, let p correspond to a marked Riemann surface S and let T be modelled on the Beltrami coefficients on S so that the zero Beltrami coefficient corresponds to p. Then q is represented by an extremal μ with $||\mu||_\infty = k$ and $d(p,q) = \frac{1}{2}\log(1+k)/(1-k)$. The existence of such an extremal μ follows from a normal families argument applied to quasiconformal mappings of bounded dilatation. For such a μ form $f(z) = [z\mu/k]$. Clearly f is holomorphic from Δ into T, $f(0) = p$, $f(k) = q$, and so the Teichmüller distance from p to q is the Poincaré distance from 0 to k. Thus, $d_K(p,q) \leq d(p,q)$. This half of the proof does not depend on the finite dimensionality of T.

For the final part of the theorem, we observe that when the infimum in (3.16) is a metric (satisfying the triangle inequality) then it is identical to the Kobayashi metric. (In the general case, to form the Kobayashi metric it is necessary to enlarge the set over which one takes infimum by taking chains of points $p = p_0, p_1, \ldots, p_n = q$ and then forming (3.16) for each pair p_{i-1}, p_i [11].)

4. *Royden's Theorem in the infinite dimensional case*

Our objective now is to show that whenever T has complex structure, even when T is infinite dimensional, that Teichmüller's and Kobayashi's metrics coincide. In view of the proof of Theorem 3.2, what we must show is that whenever $f: \Delta \longrightarrow T$ is a holomorphic mapping with $f(0) = 0$ and $f(r) = [\mu]$ where μ is extremal and $\|\mu\|_\infty = k$ and $r > 0$, then $r \geqslant k$. From Theorem 3.2 we know that this inequality holds whenever T is finite dimensional.

Now, all Teichmüller spaces with complex structure (with one exception) can be viewed as coming from Beltrami differentials in the upper half plane U. The one exception is the case of a torus, but in that case the Teichmüller space itself is isomorphic to U, and in that case it is well known that d_K and d are both identical with the Poincaré metric.

Let Γ be any Fuchsian group (possibly the identity) acting on U and let $\Lambda \subseteq \hat{\mathbf{R}}$ be its limit set. (Recall that $\hat{\mathbb{R}}$, $M(\Gamma)$, and $T(\Gamma)$ are defined in Section 1.) In the present situation we let $C = \hat{\mathbf{R}}$ so that μ and υ are equivalent if $w_\mu(x) = w_\upsilon(x)$ for all x in $\hat{\mathbf{R}}$. It is clear that there exists a sequence of finitely generated subgroups Γ_n of Γ and subsets C_n of $\hat{\mathbf{R}}$ with the following properties:

i) $\Gamma_n \subseteq \Gamma_{n+1}$ and $\bigcup \Gamma_n = \Gamma$,

ii) each Γ_n contains elements with fixed points in the intervals $I_{kn} = ((k-1)/n, k/n)$ for $-n^2 \leqslant k \leqslant n^2$ whenever $I_{kn} \cap \Lambda \neq \phi$,

iii) C_n is invariant under Γ_n, $C_n \supseteq \Lambda_n$ and $(C_n - \Lambda_n)/\Gamma_n$ is a finite set,

iv) $C_n \subseteq C_{n+1}$ and $\overline{\bigcup C_n} = \hat{\mathbb{R}}$.

Now, let $\Omega_n = \mathbb{C} \cup \{\infty\} - C_n$. We introduce a new set of Beltrami coefficients. It consists of complex-valued measurable functions μ with support in Ω_n for which $\|\mu\|_\infty < 1$ and for which

$$\mu(Az)\overline{A'(z)} = \mu(z)A'(z) \tag{4.1}$$

or all A in Γ_n. The set of such Beltrami coefficients will be denoted by (Γ_n,Ω_n).

Let $w = w^\mu$ be the unique homeomorphic self mapping of $\mathbb{C} \cup \{\infty\}$ satisfying

$$\frac{\partial w}{\partial \bar{z}} = \mu \frac{\partial}{\partial z} w \tag{4.2}$$

nd normalized to fix 0,1, and ∞. Define μ to be strongly equivalent to υ and rite $\mu \equiv_n \upsilon$ if $w^\mu(x) = w^\upsilon(x)$ for all x in C_n and if w^μ is homotopic to w^υ n Ω_n. Define $\tilde{T}(\Gamma_n,\Omega_n)$ to be $M(\Gamma_n,\Omega_n)/\equiv_n$.

Let $\pi : M(\Gamma) \longrightarrow M(\Gamma_n,\Omega_n)$ be defined by $\pi(\mu) = \mu(z)$ for z in U and $(\mu) = 0$ for z in L.

EMMA 4.1. *If* $\mu \sim \upsilon$, *then* $\pi(\mu) \equiv_n \pi(\upsilon)$.

roof. The hypothesis tells us that $w_\mu(x) = w_\upsilon(x)$ for all x in $\hat{\mathbb{R}}$, since in the ase under consideration $C = \hat{\mathbb{R}}$. This implies $w^{\pi(\mu)}(x) = w^{\pi(\upsilon)}(x)$ for all x in and hence for all x in C_n. Furthermore, by Ahlfors [1, page 119], there is homotopy $h_t : U \longrightarrow w^{\pi(\mu)}(U)$ for which $h_0(z) = w^{\pi(\mu)}(z)$ for z in U and $_1(z) = w^{\pi(\mu)}(z)$ for z in U and $h_t(x) = w^{\pi(\mu)}(x) = w^{\pi(\upsilon)}(x)$ for x in $\hat{\mathbb{R}}$ and $\leqslant t \leqslant 1$. This homotopy extends to a homotopy h_t from Ω_n to $w^{\pi(\mu)}(\Omega_n)$ by setting $_t(z) = w^{\pi(\mu)}(z) = w^{\pi(\upsilon)}(z)$ for z in $L \cup \hat{\mathbb{R}}$. It follows that $\pi(\mu)$ and $\pi(\upsilon)$ re strongly equivalent.

Lemma 4.1 implies that the mapping π induces a mapping from $T(\Gamma)$ to $\tilde{T}(\Gamma_n,\Omega_n)$. e denote this new mapping by the same letter π. Since the complex structures on (Γ) and on $\tilde{T}(\Gamma_n,\Omega_n)$ are inherited from $M(\Gamma)$ and $M(\Gamma_n,\Omega_n)$, this new mapping is olomorphic.

Now suppose μ is extremal in its class in $M(\Gamma)$. By this we mean that $= \|\mu\|_\infty \leqslant \|\upsilon\|_\infty$ for all υ in $M(\Gamma)$ for which $\upsilon \sim \mu$. Let

$$k_n|\eta_n|/\eta_n \tag{4.3}$$

be extremal in the class of $\pi(\mu)$ in $M(\Gamma_n,\Omega_n)$ under the equivalence relation $\equiv_n$. Since $\tilde{T}(\Gamma_n,\Omega_n)$ is isomorphic to an ordinary finite dimensional Teichmüller space, from Teichmüller's theorem it follows that the class of $\pi(\mu)$ in $M(\Gamma_n,\Omega_n)$ possesses a unique extremal element of the form (4.3) where $0 < k_n < 1$ and η_n is an integrable holomorphic quadratic differential on Ω_n/Γ_n except for at most simple poles at elliptic and parabolic punctures and at the points of $C_n - \Lambda_n/\Gamma_n$.

Obviously, $k_n \leq k_{n+1} \leq k$ for all n.

LEMMA 4.2. *The sequence* k_n *monotonically increases to* k.

Proof. Consider the mappings w^{ν_n} where $\nu_n = k_n|\eta_n|/\eta_n$. By hypothesis $w^{\nu_n}(x) = w^{\pi(\mu)}(x)$ for all x in C_n. Let w^{ν} be a normalized limit of some subsequence of w^{ν_n}. Such a limit exists because $\|\nu\|_\infty = k_n < k < 1$ for all n. Also $\nu(Az)\overline{A'(z)} = \nu(z)A'(z)$ for all A in Γ. From the fact that $\overline{\bigcup_n C_n} = \hat{\mathbb{R}}$ it follows that $w^{\nu}(x) = w^{\pi(\mu)}(x)$ for all x in $\hat{\mathbb{R}}$, [8]. Thus ν restricted to the upper half-plane is equivalent to μ. Moreover, ν restricted to L is trivial in $M(\Gamma,L)$ (but ν might not be identically zero in L). By the fact that μ is extremal in its class, $\|\nu|U\|_\infty \geq \|\mu\|_\infty = k$. But if $k_n \leq k-\varepsilon$ for all n and some positive ε, one would have $\|\nu\|_\infty \leq k-\varepsilon$, a contradiction. Hence, the lemma follows.

THEOREM 4.1. *For any complex Teichmüller space of a Fuchsian group, the Kobayashi and Teichmüller metrics coincide.*

Proof. From the proof of Theorem 3.2 it suffices to show that given a holomorphic f from Δ into $T(\Gamma)$ with $f(0) = 0$ and $f(r) = [\mu]$ where μ is extremal and $0 < r < 1$, then $r \geq \|\mu\|_\infty$. Now the mapping $\pi \circ f$ from Δ into $\tilde{T}(\Gamma_n,\Omega_n)$ is holomorphic and takes 0 into 0 and maps into a finite dimensional Teichmüller space. Therefore, by Theorem 3.2, $r \geq k_n$ where k_n is defined in (4.3). From Lemma 4.2 this implies $r \geq k$ and this concludes the proof of the theorem.

BROOKLYN COLLEGE, CUNY
NEW YORK, N.Y.

REFERENCES

[1] L.V. Ahlfors, *Lectures on Quasiconformal Mappings*, (Princeton, N.J.: Van Nostrand, 1966.

[2] __________, "An extension of Schwarz's lemma," *Trans. Am. Math. Soc.* 43 (1938), 359-364.

[3] L.V. Ahlfors, and L. Bers, "Riemann's mapping theorem for variable metrics," *Ann. of Math.* 72 (1960), 385-404.

[4] L. Bers, L., "A new proof of a fundamental inequality for quasiconformal mappings," *J. d'Analyse Math.* 36 (1979), 15-30.

[5] ______, "Automorphic forms and Poincaré series for infinitely generated Fuchsian groups," *Amer. J. Math.* 87 (1965), 196-214.

[6] M.M. Day, *Normed Linear Spaces*, Academic Press, New York, 1962.

[7] C.J. Earle, and J. Eells, "On the differential geometry of Teichmüller spaces," *J. Analyse Math.* 19 (1967), 35-52.

[8] F.P. Gardiner, "An analysis of the group operation in universal Teichmüller space," *Trans. Amer. Math. Soc.* 132 (1968), 471-486.

[9] ____________, "Approximation of infinite dimensional Teichmüller spaces," to appear.

[10] R.S. Hamilton, "Extremal quasiconformal mappings with prescribed boundary values," *Trans. Amer. Math. Soc.* 138 (1969), 399-406.

[11] S. Kobayashi, *Hyperbolic Manifolds and Holomorphic Mappings*, Marcel Dekker, Inc. N.Y., 1970.

[12] B. O'Byrne, "On Finsler geometry and applications to Teichmüller spaces," (Ahlfors et al., ed.) *Ann. of Math. Studies* 66 (1971), 317-328.

[13] E. Reich and K. Strebel, "Extremal quasiconformal mappings with given boundary values," in *Contributions to Analysis*, 375-392, ed. L.V. Ahlfors et al. (New York and London: Academic Press, 1974).

[14] H. Royden, "Automorphisms and isometries of Teichmüller spaces," (Ahlfors et al, ed.) *Ann. of Math. Studies* 66 (1971), 365-367.

[15] K. Strebel, "On quasiconformal mappings of open Riemann surfaces," *Comment. Math. Helv.* 53 (1978), 301-321.

[16] _________, "On the trajectory structure of quadratic differentials," (Greenberg ed.) *Ann. of Math. Studies* 79 (1974), 419-438.

THE ELEMENTARY THEORY OF CORRESPONDENCES

GEORGE R. KEMPF*

Introduction

In this paper we will revisit some of the former glory of the theory of correspondences between algebraic curves. This classical topic has been discussed by all important schools of algebraic geometry in the past. Here I have reworked some of the results from A. Weil's monograph [1] using the sheaf theory. The "elementary" in the title refers to the fact that the presentation uses one-dimensional methods as opposed to using the theory of surfaces or Jacobian varieties in an essential way.

1. *Raw material about correspondences*

Let $\mathcal{L}$ be an invertible sheaf on the product $C \times D$ of two complete smooth curves. Then $\mathcal{L}$ will be called a *correspondence* from D to C. A correspondence $\mathcal{L}$ is trivial if it is isomorphic to a sheaf of the form $\pi_D^* \mathcal{M} \otimes_{\mathcal{O}_{C \times D}} \pi_C^* \mathcal{N}$, where $\mathcal{M}$ and $\mathcal{N}$ are invertible sheaves on D and C.

Every theory of correspondences gives a numerical criterion for a correspondence to be trivial. Before I develop such a criterion, I want to explain a very elementary idea, which shows that some correspondences are trivial.

LEMMA 1.1. *Let* $\mathcal{L}$ *be an invertible sheaf on* $C \times D$ *and* $\mathcal{M}$ *an invertible sheaf on* D. *Assume that we have an* $\mathcal{O}_{C \times D}$*-homomorphism* $\psi : \pi_D^* \mathcal{M} \to \mathcal{L}$ *such that for some point* c *of* C, $\psi|_{c \times D} : \mathcal{M} \to \mathcal{L}|_{c \times D}$ *is an isomorphism. Then* $\mathcal{L}$ *is a trivial correspondence of the form* $\pi_D^* \mathcal{M} \otimes \pi_C^* \mathcal{N}$ *where* $\mathcal{N}$ *is an invertible sheaf on* C.

Proof. Let R be the divisor of zeroes of ψ. Then R is an effect divisor on $C \times D$ such that $\mathcal{L} \simeq (\pi_D^* \mathcal{M})(R)$. If we can show that $R = \pi_C^{-1}$ where S is a divisor on C, we will have the required isomorphism $\mathcal{L} \approx \pi_D^* \mathcal{M} \otimes \pi_C^* \mathcal{N}$ where $\mathcal{N} = \mathcal{O}_C(S)$. Hence the lemma will be proven.

* Supported in part by NSF grant MCS77-18723(A04) and grant 7900965.

Let R_i be a component of R. By our assumption R_i does not meet the divisor $c \times D$. Thus the projection π_C maps the complete curve R_i into the incomplete curve $C - \{c\}$. Hence this projection takes a constant value c_i where $S = \Sigma c_i$. Q.E.D.

As an intermediate step toward the application of this last idea, we have the

LEMMA 1.2. Let $\mathcal{L}$ *be an invertible sheaf on* $C \times D$. *Assume that*

a) $\pi_{D*}\mathcal{L}$ *is zero and*

b) there is a point c *of* C *such that the homomorphism*
$\gamma : R^1\pi_{D*}\mathcal{L} \to R^1\pi_{D*}(\mathcal{L}(c+D))$ *is injective.*

Then $\mathcal{L}$ *is a trivial correspondence.*

Proof. We have a short exact sequence,

$$0 \to \mathcal{L} \to \mathcal{L}(c \times D) \to \mathcal{L}(c \times D)|_{c \times D} \to 0.$$

Taking its direct images, we have a long exact sequence,

$$0 \to \pi_{D*}\mathcal{L} \overset{\alpha}{\to} \pi_{D*}(\mathcal{L}(c\times D)) \overset{\beta}{\to} \pi_{D*}(\mathcal{L}(c\times D)|_{c\times D}) \overset{\delta}{\to} R^1\pi_{D*}\mathcal{L} \overset{\gamma}{\to} R^1\pi_{D*}(\mathcal{L}(c\times D)).$$

Our assumptions tell us that α and δ are zero. Thus, β is an isomorphism where $\mathcal{M} \equiv \pi_{D*}(\mathcal{L}(c \times D)|_{c \times D})$ is an invertible sheaf on D. Associated to the inverse $\beta^{-1} : \mathcal{M} \to \pi_{D*}(\mathcal{L}(c \times D))$, we have an $\mathcal{O}_{C \times D}$-homomorphism

$$\psi : \pi_D{}^*\mathcal{M} \longrightarrow \mathcal{L}(c \times D), \quad \text{such that}$$

$$\psi|_{c \times D} : \mathcal{M} \to \mathcal{L}(c \times D)|_{c \times D} \quad \text{is an isomorphism.}$$

Therefore, by the preceding lemma, $\mathcal{L}(c \times D)$ and, hence, $\mathcal{L}$ itself are trivial correspondences. Q.E.D.

Next we need to know the two partial degrees of a correspondence between C and D. Define $\deg_C \mathcal{L}$ to be the degree of the invertible $\mathcal{O}_C$-module $\equiv \mathcal{L}|_{C \times d}$ for any or all points d of D. Similarly $\deg_D \mathcal{L} \equiv \deg(\mathcal{L}|_{c \times D})$ for any c in C.

Another step toward our criterion for triviality is

LEMMA 1.3. Let $\mathcal{L}$ *be an invertible sheaf on* $C \times D$. *Assume that* $\deg_C \mathcal{L} = \text{genus}(C) - 1$ *and that* $\Gamma(C, \mathcal{L}|_{C \times e})$ *is zero for one point* e *of* D. *Then,*

a) $\pi_{D*} \mathcal{L}$ *is zero,*

b) $R^1 \pi_{D*} \mathcal{L}$ *is a torsion coherent sheaf on* D, *and*

c) $\mathcal{L}$ *is a trivial correspondence* $\Longleftrightarrow$ $R^1 \pi_{D*} \mathcal{L}$ *is zero.*

Proof. As $\Gamma(C, \mathcal{L}|_{C \times e}) = 0$, $\Gamma(C, \mathcal{L}|_{C \times d}) = 0$ for all points d of some open dense neighborhood U of e by upper-semi-continuity. Thus $\Gamma(C \times W, \mathcal{L}) = 0$ for any open subset W of C. Hence $\pi_{D*} \mathcal{L} = 0$; i.e., a) is true. For b), by our assumption on $\deg_C \mathcal{L}$, $\dim \Gamma(C, \mathcal{L}|_{C \times d}) = \dim H^1(C, \mathcal{L}|_{C \times d})$ for any point d of D. Thus $H^1(C, \mathcal{L}|_{C \times d}) = 0$ when d is a point of U. Consequently the restriction of $R^1 \pi_{D*} \mathcal{L}$ to U is zero and, hence, $R^1 \pi_{D*} \mathcal{L}$ is a torsion sheaf on D. Therefore because $R^1 \pi_{D*} \mathcal{L}$ is coherent, b) is also true.

For c), assume that $\mathcal{L}$ is a trivial correspondence, then all the sheaves $\mathcal{L}|_{C \times d}$ are isomorphic. Hence we may take the above open set U to be all of D. Consequently as before we have $R^1 \pi_{D*} \mathcal{L} = 0$. Conversely assume that $R^1 \pi_{D*} \mathcal{L} = 0$. Then the assumptions of Lemma 1.2 are verified for any point c of C. Thus that lemma implies that $\mathcal{L}$ is a trivial correspondence. Therefore c) is true. Q.E.D.

Next we will give a global interpretation of this result.

LEMMA 1.4. In the situation of Lemma 1.3,

a) the only non-zero cohomology group of $\mathcal{L}$ *is* $H^1(C \times D, \mathcal{L})$ *which is naturally isomorphic to* $\Gamma(D, R^1\pi_{D*}\mathcal{L})$, *and*

b) $\mathcal{L}$ *is a trivial correspondence* $\Longleftrightarrow H^1(C \times D, \mathcal{L})$ *is zero.*

Proof. As $R^i\pi_{D*}\mathcal{L} = 0$ if $i > 1 = \dim C$, $R^1\pi_{D*}\mathcal{L}$ is the only non-zero higher direct image of $\mathcal{L}$ by Lemma 1.3a). Thus the Leray spectral sequence for the projection π_D gives natural isomorphisms $H^i(C \times D, \mathcal{L}) \approx H^{i-1}(D, R^1\pi_{D*}\mathcal{L})$ for all i. By Lemma 1.3b), $R^1\pi_{D*}\mathcal{L}$ has no higher cohomology because its support consists of a finite number of points and, hence, a) follows from the above isomorphism. Furthermore Lemma 1.3b) also implies that $R^1\pi_{D*}\mathcal{L} = 0 \Longleftrightarrow \Gamma(D, R^1\pi_{D*}\mathcal{L}) = 0$. Therefore b) follows from a) together with Lemma 1.3c). Q.E.D.

Using the Euler characteristic $\chi(\mathfrak{F}) = \Sigma(-1)^i \dim H^i(X, \mathfrak{F})$ of a coherent sheaf $\mathfrak{F}$ on a complete variety X, we may deduce immediately the

COROLLARY 1.5. In the situation of Lemma 1.3,

a) $\chi(\mathcal{L}) \leq 0$, *and*

b) $\mathcal{L}$ *is a trivial correspondence* $\Longleftrightarrow \chi(\mathcal{L}) = 0$.

In the next section we will explain how to modify this result so that it applies to arbitrary correspondences without the restriction $\deg_C \mathcal{L} = \text{genus}(C) - 1$.

2. *The numerical function on correspondences*

To make a numerical measure of how much a correspondence $\mathcal{L}$ from D to C is twisted up (i.e., non-trivial), consider the expression

$$N(\mathcal{L}) \equiv -\chi(\mathcal{L}) + \chi_D(\mathcal{L}) \cdot \chi_C(\mathcal{L}),$$

where $\chi_D(\mathcal{L}) \equiv \deg_D(\mathcal{L}) + \chi(\mathcal{O}_D)$ and $\chi_C(\mathcal{L}) \equiv \deg_C(\mathcal{L}) + \chi(\mathcal{O}_C)$. By the Künneth formula, this expression $N(\mathcal{L})$ may be seen to vanish when $\mathcal{L}$ is a trivial correspondence but we will not need to use this idea directly.

We will first note how the numerical function changes as we make a slight change in its variable.

LEMMA 2.1. *Let* E *be a effective divisor on* $C \times D$ *and let* $\mathcal{L}$ *be a correspondence from* D *to* C. *Let* $\mathcal{M} = \mathcal{O}_{C \times D}(-E)$. *Then*

$$N(\mathcal{L}\otimes\mathcal{M}) = N(\mathcal{L}) + \chi(\mathcal{L}|_E) + \deg_D\mathcal{M} \cdot \chi_C(\mathcal{L}) + \chi_D(\mathcal{L}) \cdot \deg_C\mathcal{M} + \deg_C\mathcal{M} \cdot \deg_D\mathcal{M} .$$

or, equivalently,

$$N(\mathcal{L}\otimes\mathcal{M}) = N(\mathcal{L}) + \deg(\mathcal{L}|_E) + \deg_C\mathcal{L} \cdot \deg_D\mathcal{M} + \deg_D\mathcal{L} \cdot \deg_C\mathcal{M}$$
$$+ \chi(\mathcal{O}_E) + \deg_D\mathcal{M} \cdot \chi(\mathcal{O}_C) + \chi(\mathcal{O}_D) \cdot \deg_C\mathcal{M} + \deg_C\mathcal{M}\,\deg_D\mathcal{M} .$$

Proof. We have an exact sequence $0 \to \mathcal{L}(-E) \to \mathcal{L} \to \mathcal{L}|_E \to 0$ of sheaves on $C \times D$. Taking Euler characteristics, we get that $\chi(\mathcal{L} \otimes \mathcal{M}) = \chi(\mathcal{L}) - \chi(\mathcal{L}|_E)$. As $\chi_*(\mathcal{L} \otimes \mathcal{M}) = \chi_*(\mathcal{L}) + \deg_*\mathcal{M}$ when $* = C$ or D, we immediately deduce that $\chi_D(\mathcal{L} \otimes \mathcal{M}) \cdot \chi_C(\mathcal{L} \otimes \mathcal{M}) = \chi_D(\mathcal{L}) \cdot \chi_C(\mathcal{L}) + \chi_D(\mathcal{L})\deg_C\mathcal{M} + \chi_C(\mathcal{L})\deg_D\mathcal{M} + \deg_C\mathcal{M} \cdot \deg_D\mathcal{M}$. The first equation in the lemma follows directly from the above two equations and the definition of the numerical function N. The second equivalent form follows from the definition of $\chi_*(\mathcal{L})$ and the equation $\chi(\mathcal{L}|_E) = \deg(\mathcal{L}|_E) + \chi(\mathcal{O}_E)$. Q.E.D.

Two correspondences $\mathcal{L}_1$ and $\mathcal{L}_2$ from D to C are called equivalent if $\mathcal{L}_1 \otimes \mathcal{L}_2^{\otimes -1}$ is a trivial correspondence. This notion defines equivalence classes which are called correspondence classes.

As an immediate consequence of the last lemma, we have

PROPOSITION 2.2. The numerical function $N(\mathcal{L})$ *is constant on correspondence classes.*

Proof. Let $E = c \times D$ for any point c of C. Then $\chi(\mathcal{L}|_E) = \chi_D(\mathcal{L})$ and, if $\mathcal{M} = \mathcal{O}_{C \times D}(-E)$, $\deg_C \mathcal{M} = -1$ and $\deg_D \mathcal{M} = 0$. Immediately from the first equation in Lemma 2.1, we have $N(\mathcal{L} \otimes \mathcal{M}) = N(\mathcal{L})$. Therefore by induction $N(\mathcal{L}) = N(\mathcal{L} \otimes \pi_C^* \mathcal{N})$ for any invertible sheaf $\mathcal{N}$ on C. Thus by symmetry between C and D, we see that $N(\mathcal{L})$ is constant on correspondence classes. Q.E.D.

We are ready to state the main result.

THEOREM 2.3. Let $\mathcal{L}$ *be a correspondence from* D *to* C. *Then,*

a) $N(\mathcal{L}) \geq 0$ *and*

b) $N(\mathcal{L}) = 0 \Longleftrightarrow \mathcal{L}$ *is a trivial correspondence.*

Proof. By Proposition 2.2 we need only show that the theorem is true for a carefully chosen representative of any correspondence class. Let d be a fixed point of D and $\mathcal{L}$ be the given correspondence from D to C. By adding or subtracting divisors of the form $c \times D$ from $\mathcal{L}$, we may assume that $\deg_C(\mathcal{L}) = \text{genus}(C) - 1 \equiv -\chi(\mathcal{O}_C)$. If $\Gamma(C, \mathcal{L}|_{C \times d}) = 0$, we are finished by Corollary 1.5 because $N(\mathcal{L}) = -\chi(\mathcal{L})$ as $\chi_C(\mathcal{L}) = 0$.

Otherwise, for general points c_1 and c_2 of C, $\dim \Gamma(C, \mathcal{L}(c_1 \times D - c_2 \times D)|_{C \times d}) < \dim \Gamma(C, \mathcal{L}|_{C \times d})$. Hence, by induction, any correspondence is equivalent to a correspondence for which the theorem has been proven. Q.E.D.

With our previous results we may deduce

COROLLARY 2.4. *Let* $\mathcal{L}$ *be a correspondence from* D *to* C *which is equivalent to a correspondence of the form* $\mathcal{O}_{C\times D}(-E)$, *where* E *is an effective divisor on* $C \times D$. *Then*

$$N(\mathcal{L}) = \chi(\mathcal{O}_E) + \deg_D \mathcal{L} \cdot \chi(\mathcal{O}_C) + \chi(\mathcal{O}_D) \cdot \deg_C \mathcal{L} + \deg_C \mathcal{L} \cdot \deg_D \mathcal{L} .$$

Proof. By Proposition 2.2 we need only prove this when $\mathcal{L} \approx \mathcal{O}_{C\times D}(-E)$. To get this formula one applies the second formula of Lemma 2.1 with $\mathcal{L}$ equal $\mathcal{O}_{C\times D}$ and $\mathcal{M}$ equal the present $\mathcal{L}$ and then notes that $N(\mathcal{O}_{C\times D}) = 0$ by Theorem 2.3b). Q.E.D.

One pleasant fact about correspondences is that the hypothesis of the last corollary is always verified.

LEMMA 2.5. *Any correspondence* $\mathcal{L}$ *from* D *to* C *is equivalent to a correspondence of the form* $\mathcal{O}_{C\times D}(-E)$ *where* E *is an effective diviso- on* $C \times D$.

Proof. Take any points c and d on C and D. Then $H = c \times D + C \times d$ is an ample divisor on the product $C \times D$. Thus the sheaf $\mathcal{L}^{\otimes -1}(mH)$ has many non-zero sections when $m \to \infty$. We may take the desired divisor E to be the zero divisor of any of the above sections. Q.E.D.

3. *More about the numerical function* $N(\mathcal{L})$

We have seen in the last section that the number $N(\mathcal{L})$ only depends on the correspondence class of $\mathcal{L}$. In this section we will make a finer study of this function. We will begin with the following fact, which expresses duality in one form or other.

LEMMA 3.1. For any correspondence $\mathcal{L}$ *from* D *to* C, *we have* $N(\mathcal{L}) = N(\mathcal{L}^{\otimes -1})$.

Proof. By Lemma 2.5 and Proposition 2.2, we may assume that $\mathcal{L} \approx \mathcal{O}_{C\times D}(-E)$ where E is an effective divisor on $C \times D$. As the dualizing sheaf $\omega = \Lambda^2 \Omega^1_{C\times D}$ is just $\pi_C^* \Omega_C \otimes \pi_D^* \Omega_D$, ω is a trivial correspondence. Hence $N(\omega) = 0$ by Theorem 2.3b) and $N(\mathcal{L}^{\otimes -1}) = N(\omega \otimes \mathcal{L}^{\otimes -1})$ by Proposition 2.2. If we use the first equation of Lemma 2.1 with $\mathcal{L}$ equal to the present $\omega \otimes \mathcal{L}^{\otimes -1}$ and $\mathcal{M}$ equal to the present $\mathcal{L}$ together with the last two equations, we may deduce that

$$N(\mathcal{L}^{\otimes -1}) = -(\omega \otimes \mathcal{L}^{\otimes -1}|_E) - \deg_D \mathcal{L} \cdot \chi_C(\omega \otimes \mathcal{L}^{\otimes -1}) - \deg_C \mathcal{L} \cdot \chi_D(\omega \otimes \mathcal{L}^{\otimes -1})$$
$$-\deg_C \mathcal{L} \deg_D \mathcal{L}.$$

By the adjunction formula $\omega \otimes \mathcal{L}^{\otimes -1}|_E \approx \omega(E)|_E$ is isomorphic to the dualizing sheaf on the curve E. Thus $-\chi(\omega \otimes \mathcal{L}^{\otimes -1}|_E) = \chi(\mathcal{O}_E)$. Furthermore, for any point d on D, the sheaf $\omega|_{C\times d}$ is isomorphic to the dualizing sheaf Ω_C on C. Therefore $-\chi_C(\omega \otimes \mathcal{L}^{\otimes -1}) = -\chi(\Omega_C) + \deg_C \mathcal{L} = \chi(\mathcal{O}_C) + \deg_C \mathcal{L}$. Similarly $-\chi_D(\omega \otimes \mathcal{L}^{\otimes -1}) = \chi(\mathcal{O}_D) + \deg_C \mathcal{L}$. If we use these three equations together with the last equation in the last paragraph, we get that

$$N(\mathcal{L}^{\otimes -1}) = \chi(\mathcal{O}_E) + \deg_D \mathcal{L} \cdot \chi(\mathcal{O}_C) + \deg_C \mathcal{L} \cdot \chi(\mathcal{O}_D) + \deg_C \mathcal{L} \cdot \deg_D \mathcal{L},$$

which equals $N(\mathcal{L})$ by Corollary 2.4. Q.E.D.

Another important aspect of our numerical function may be expressed in terms of an inner product. Let $\mathcal{L}$ and $\mathcal{M}$ be two correspondences from D to C. We may form the expression

$$\langle \mathcal{L}, \mathcal{M} \rangle \equiv N(\mathcal{L} \otimes \mathcal{M}) - N(\mathcal{L}) - N(\mathcal{M}) + N(\mathcal{O}_{C\times D}).$$

This expression has some simple properties.

PROPOSITION 3.2. a) The expression $<\mathcal{L}, \mathcal{M}>$ *is symmetric, bi-additive and depends only on the correspondence classes of* $\mathcal{L}$ *and* $\mathcal{M}$.

b) If $\mathcal{M}$ *is equivalent to a correspondence of the form* $\mathcal{O}_{C \times D}(-E)$ *for some effective divisor* E *on* $C \times D$, *then* $<\mathcal{L}, \mathcal{M}> = \deg(\mathcal{L}|_E) + \deg_C \mathcal{L} \cdot \deg_D \mathcal{M} + \deg_D \mathcal{L} \cdot \deg_C \mathcal{M}$.

Proof. Clearly $<\mathcal{L}, \mathcal{M}> = <\mathcal{M}, \mathcal{L}>$ by definition and thus the expression is symmetric. Also, as the numerical function N is a correspondence class invariant, the inner product is also. Furthermore by the symmetry of the inner product, to verify the bi-additivity we need to show that $<\mathcal{L}_1 \otimes \mathcal{L}_2^{\otimes \pm 1}, \mathcal{M}> = <\mathcal{L}_1, \mathcal{M}> \pm <\mathcal{L}_2, \mathcal{M}>$. This equation is implied by the formula of part b) since we may use Lemma 2.5 to verify the hypothesis on $\mathcal{M}$. Thus b) implies a).

To check b) one only has to use the second equation in Lemma 2.1 and the same equation with $\mathcal{L}$ equal to $\mathcal{O}_{C \times D}$. The necessary calculation may be done on sight from the definition of the inner product. Q.E.D.

Finally we may use our inner product to show that the numerical function is quadratic. When $\mathcal{L} = \mathcal{O}_{C \times D}(F)$, the degree $\deg_{C \times D}(\mathcal{L})$ is defined as the self intersection number of F.

PROPOSITION 3.3. For any correspondence $\mathcal{L}$ *from* D *to* C,

a) $N(\mathcal{L}) = \frac{1}{2} <\mathcal{L}, \mathcal{L}>$, *and*

b) $N(\mathcal{L}) = -\frac{1}{2} \deg_{C \times D}(\mathcal{L}) + \deg_C \mathcal{L} \cdot \deg_D \mathcal{L}$.

Proof. Consider the inner product $<\mathcal{L}, \mathcal{L}^{\otimes -1}>$. By Proposition 3.2a) it equals $-<\mathcal{L}, \mathcal{L}>$. On the other hand by definition it also equals

$N(\mathcal{L} \otimes \mathcal{L}^{\otimes -1}) - N(\mathcal{L}) - N(\mathcal{L}^{\otimes -1}) + N(\mathcal{O}_{C \times D})$. By Lemma 3.1 and Theorem 2.3b) the last expression is just $-2N(\mathcal{L})$. Thus $-\langle \mathcal{L}, \mathcal{L} \rangle = -2N(\mathcal{L})$ and, hence, a) is true.

For b), if $\mathcal{L} \approx \mathcal{O}_{C \times D}(-E)$, with E an effective divisor on $C \times D$, then $\frac{1}{2} \langle \mathcal{L}, \mathcal{L} \rangle = \deg(\mathcal{L}|_E) + 2\deg_C \mathcal{L} \cdot \deg_D \mathcal{L}$ by Proposition 3.2b). As $\deg(\mathcal{L}|_E) = -\deg_{C \times D}(\mathcal{L})$, the formula in b) is true in this case. One may easily check that the right side of the formula in b) is a correspondence invariant. Thus b) is true in general by Theorem 2.3a) and Lemma 2.5. Q.E.D.

Let $\mathcal{L}$ and $\mathcal{M}$ be two correspondences from D to C. By using Propositions 3.2a) and 3.3a), for any integers ℓ and n,

$$*) \qquad N(\mathcal{L}^{\otimes \ell} \otimes \mathcal{M}^{\otimes m}) = N(\mathcal{L}) \cdot \ell^2 + \langle \mathcal{L}, \mathcal{M} \rangle \ell \cdot m + N(\mathcal{M}) \cdot m^2.$$

Thus the above expression is a quadratic function of (ℓ, m), which is positive semi-definite by Theorem 2.3 a). Therefore its discriminant must be non-positive; i.e.,

$$\langle \mathcal{L}, \mathcal{M} \rangle^2 - 4N(\mathcal{L})N(\mathcal{M}) \le 0.$$

Writing this inequality another way, we have proven

PROPOSITION 3.4. $|\langle \mathcal{L}, \mathcal{M} \rangle| \le 2\sqrt{N(\mathcal{L})N(\mathcal{M})}$.

Frequently in applications one uses the above inequality for the correspondences which arise from morphisms from D to C. Let $\lambda : D \to C$ be a morphism and let $\Gamma_\lambda \subset C \times D$ be its graph. Then $\mathcal{L}_\lambda \equiv \mathcal{O}_{C \times D}(\Gamma_\lambda)$ is the correspondence associated to the morphism. Also for two distinct morphisms λ and μ from D to C the number $[\lambda, \mu]$ of coincidences between λ and μ is defined to be the intersec tion number $[\Gamma_\lambda \cdot \Gamma_\mu]$. With these notations, we have

LEMMA 3.5. a) $N(\mathcal{L}_\lambda) = \text{genus}(C) \cdot \deg \lambda$.

b) $\langle \mathcal{L}_\lambda, \mathcal{L}_\mu \rangle = \deg \lambda + \deg \mu - [\lambda, \mu]$

Proof. By Lemma 3.1 and Corollary 2.4, $N(\mathcal{L}_\lambda) = N(\mathcal{L}_\lambda^{\otimes -1})$ $= \chi(\mathcal{O}_{\Gamma_\lambda}) - \deg_D \mathcal{L}_\lambda \cdot \chi(\mathcal{O}_C) - \chi(\mathcal{O}_D) \cdot \deg_C \mathcal{L}_\lambda + \deg_C \mathcal{L}_\lambda \cdot \deg_D \mathcal{L}_\lambda$. As Γ_λ is the graph of a morphism $\lambda : D \to C$, $\Gamma \approx D$, $\deg_D \mathcal{L}_\lambda = \deg \lambda$ and $\deg_C \mathcal{L}_\lambda = 1$. Thus $N(\mathcal{L}_\lambda) = \chi(\mathcal{O}_D) - \deg\lambda \cdot \chi(\mathcal{O}_C) - \chi(\mathcal{O}_D) +$ $\deg \lambda = \deg\lambda[1 - \chi(\mathcal{O}_C)] = \deg\lambda \cdot \text{genus}(C)$. Thus a) is true. For b) by Proposition 3.2, $-\langle \mathcal{L}_\lambda, \mathcal{L}_\mu \rangle = \langle \mathcal{L}_\lambda, \mathcal{L}_\mu^{\otimes -1} \rangle = \deg(\mathcal{L}_\lambda |_{\Gamma_\mu}) + 1 \cdot (-\deg\mu) +$ $(\deg\lambda)(-1)$. As $\deg(\mathcal{L}_\lambda |_{\Gamma_\mu}) = [\Gamma_\lambda \cdot \Gamma_\mu] = [\lambda, \mu]$, $-\langle \mathcal{L}_\lambda, \mathcal{L}_\mu \rangle = [\lambda, \mu] -$ $\deg \mu - \deg \lambda$ and b) is true. Q.E.D.

THE JOHNS HOPKINS UNIVERSITY AND
THE INSTITUTE FOR ADVANCED STUDY

REFERENCE

1. A. Weil, *Sur les courbes algébriques et les variétés qui s'en déduisent*, Paris, Hermann, 1948.

PANELLED WEB GROUPS

BERNARD MASKIT*

The main object of this paper is to give a more or less explicit construction and description of a certain class of Kleinian groups. The corresponding class of hyperbolic 3-manifolds can be described as having incompressibly embedded boundary, and having in some sense a maximal number of essential annuli.

If we drop the requirement that the boundary components be incompressibly embedded, then we can regard our class as being the class of Kleinian groups that can be built up from elementary groups using combination theorems, where the amalgamated or conjugated subgroup is cyclic. One can describe all such groups using the description of function groups (i.e., groups which keep invariant some region of the extended complex plane) [8]. The description of function groups is already quite complicated, and describing the groups in this new class is even more so. We shortcut these difficulties here by dealing only with Kleinian groups that are purely loxodromic (including hyperbolic), and by describing only those Kleinian groups for which every component (of the set of discontinuity) is simply connected; i.e., web groups (this is equivalent to the requirement that the boundaries of the 3-manifold be incompressibly embedded).

Combination theorems never lead us out of the class of geometrically finite groups; all the groups constructed here are geometrically finite.

In §1, we use only combinatorial topology and construct a class of 3-manifolds. The building blocks for the construction are the 3-manifolds given by Fuchsian and extended Fuchsian groups (i.e., Kleinian groups which contain Fuchsian subgroups of index 2) of the second kind; these Kleinian groups are described in §2-3. These building blocks are put together by gluing along cylinders in the boundary; this is the combinatorial analog of the usual use of the combination theorems. The Kleinian

*Research supported in part by NSF Grant #MCS 8102621.

group use and description of these combination theorems is given in §5-6. The final operation is a variant of Dehn surgery: a solid torus is cut along a cylinder, and this cylinder is glued to a cylinder on the boundary of the previously constructed 3-manifold. The corresponding Kleinian group description involving combination theorems is given in §7.

In §4-7 we construct a particular Kleinian group whose 3-manifold is a particular one in our class. In §8 we describe how to proceed in general to construct a Kleinian group whose 3-manifold is any given member of this class.

In the last section, we define panelled web groups in terms of intersections of component subgroups of Kleinian groups, and indicate that the class of Kleinian groups constructed in §4-8, with one minor restriction, is the class of panelled web groups.

All the material here is descriptive. Proofs are either omitted entirely or are given only in broad outline. A full development of the theory of panelled groups will appear elsewhere.

1. 3-manifolds

In this section we construct a class of 3-manifolds by starting with a finite collection of I-bundles over compact surfaces, gluing them together in a specific fashion, and then using a finite number of specific operations similar to Dehn surgery.

1.1. Let S be a compact orientable surface with boundary. To be specific, we assume that S has genus p and n boundary components. We will always assume that if $p = 0$, then $n \geq 3$; otherwise there are no restrictions on p and n.

Let I be the closed unit interval $[0,1]$.

Let P be an I-bundle over S of one of the following two types:

i) $P = S \times I$.

ii) There is a freely acting orientation reversing involutory homeomorphism

$h: S \longrightarrow S$; we extend h to an orientation preserving homeomorphism $h': S \times I \longrightarrow S \times I$ by $h'(x,t) = (h(x),1-t)$. Then $P = S \times I/h'$.

Note that since h acts freely and reverses orientation, h cannot preserve any boundary component of S; in particular, we can have an I-bundle of type (ii) only if S has an even number of boundary components.

1.2. The I-bundle P has a set of *distinguished boundary curves* as follows. If W is a boundary curve of S, then $W \times \{0\}$ and $W \times \{1\}$ are the corresponding distinguished boundary curves of P; if P is of type (ii), then the distinguished boundary curves can also be regarded as just the curves $W \times \{0\}$.

The distinguished boundary curves of P are *paired*: $W \times \{0\}$ is paired with $W \times \{1\}$. This pairing is called the I-*pairing*.

1.3. In addition to the distinguished boundary curves, P has a *favored boundary surface* S'. If P is of type (i), then S' is the disjoint union of $S \times \{0\}$ and $S \times \{1\}$. If P is of type (ii), then $S' = S \times \{0\}$.

We observe that the distinguished boundary curves are precisely the boundary curves of the favored boundary.

There also are *boundary cylinders* on the boundary of P. These are the cylinders $W \times I$, where $W \times \{0\}$ is a distinguished boundary curve. Each boundary cylinder has a *median* $W \times \{\frac{1}{2}\}$ on it; the median divides the boundary cylinder into two *half-cylinders*

1.4. We start with a finite number of surfaces $S_1,\ldots,S_s$ and corresponding I-bundles as above $P_1,\ldots,P_s$, subject to two conditions. The first condition, which we have already stated, is that no S_i is either a disc or a cylinder. The second condition is that each P_i should have a positive number of distinguished boundary curves, hence also a positive number of boundary cylinders. Of course each P_i has an even number of distinguished boundary curves.

1.5. We now introduce a new pairing of the distinguished boundary curves of the

P_i. This new pairing, called the S-*pairing*, is chosen arbitrarily, but subject to two conditions. The first condition is that no distinguished boundary curve is paired with itself. The second condition is as follows: if we identify two distinguished boundary curves which are S-paired, then $P_1 \cup \cdots \cup P_s$ modulo this identification should be connected.

1.6. We now glue together the P_i using the S-pairing as a guide. The result will be a compact connected 3-manifold M with boundary. The boundary of M is the union of the favored boundaries of the P_i with their boundary loops identified by the S-pairing. The 3-manifold also contains in its interior a finite number of special loops called *medians*.

We start our construction with some distinguished boundary curve W_1' lying on the boundary of an I-bundle, say P_1. Let W_2 be the distinguished boundary curve which is S-paired with W_1'. Let A_1 be the boundary cylinder containing W_1' on ∂P_1, and let A_2 be the boundary cylinder containing W_2. We want to glue the half-cylinder of A_1 between W_1' and the median to the half-cylinder of A_2 between W_2 and the median, so that the favored boundaries near W_1' and W_2 fit together to give a surface. We require the medians to be glued together so that after gluing, we still have a 3-manifold with boundary. It is clear that this can be done.

After performing the above operation, our 3-manifold is either P_1 with two half-cylinders glued together, or say $P_1 \cup P_2$ with a half-cylinder of P_1 glued to a half-cylinder of P_2. In either case, our new 3-manifold has a new boundary cylinder in addition to the old boundary cylinders other than A_1 and A_2. The medians of A_1 and A_2 have been identified and our new boundary cylinder is the union of the unidentified half-cylinders of A_1 and A_2.

We continue our construction by letting W_2' be the distinguished boundary curve which is I-paired with W_2, and by letting W_3 be the distinguished boundary curve which is S-paired with W_2'. We let A_3 be the boundary cylinder on the boundary of the appropriate P_i containing W_3, and as above, we glue the half-cylinder containing W_2' (this half-cylinder lies on our new boundary cylinder) to the half-cylinder of A_3 containing W_3.

After the above operation we have a new 3-manifold, which no longer has the boundary cylinders A_1, A_2, A_3 but instead has one new boundary cylinder, one half of which is a half-cylinder of A_1, the other a half-cylinder of A_3.

After a finite number of steps as above, perhaps even at the first step, after gluing W_1' to W_2, W_2' to $W_3, \ldots, W_{m-1}'$ to W_m, we reach the point where W_m', the distinguished boundary curve I-paired with W_m, is S-paired with W_1, the distinguished boundary curve I-paired with W_1'. At this point the boundary cylinders $A_1, \ldots, A_m$ have all been replaced by a single boundary cylinder, call it A. The medians of the A_i have all been identified; this is the median of A. One half of A has W_1 on its boundary, the other half has W_m' on its boundary. We simply glue these two halves together, where we glue W_1 to W_m', and our gluing is the identity on the median.

1.7. After performing the above sequence of operations, there may still be boundary cylinders on the boundary of our new 3-manifold. We repeat the above sequence of operations until we have used up all the boundary cylinders.

At the end of all these operations, we are left with a compact connected 3-manifold M, where the boundary of M is the union of the favored boundaries of the P_i, with S-paired distinguished boundary curves glued together. The 3-manifold M has in its interior a finite number of disjoint loops, the medians. Each median is freely homotopic to at least one loop on the boundary of M; in general, each median is freely homotopic to a finite number of homotopically distinct loops on ∂M.

1.8. Our last operation is a form of Dehn surgery performed along some or all of the medians. For each median, this operation depends on a rational number p/q called the *complex twist*, $0 < p/q < 1$ (for convenience, we will write $p/q = 0$ to mean that we do not perform this operation).

We start with some median C. Along with C, we have the half-cylinders lying between C and the loops W_i on ∂M. We choose one of these half-cylinders (it might as well be the last one we glued together) and cut M open along it, giving us again a boundary cylinder A on ∂M. We call the boundary loops of A, W_1 and W_2, and we call the median C.

We started by asserting that none of our building blocks was of the form $S \times I$ where S is a cylinder. We now let P be such a 3-manifold which we think of as a handlebody of genus 1. We choose a simple loop C' in the interior of P, where C' generates $\pi_1(P)$; we call C' the *median* of P.

We chose a standard homotopy basis X,Y on ∂P, where X is homotopic to C' and Y is null-homotopic in P. There is a simple loop W_1' on ∂P homotopic to $X^q Y^p$. In P, W_1' is homotopic to $(C')^q$ and we can find a cylinder A', embedded in P except along C', which realizes this homotopy. We cut P along A' giving us a new boundary cylinder on ∂P, where this cylinder has $(C')^q$ as median and loops W_1', W_2' on the boundary. We glue P cut along A' to M cut along A, gluing A' to A, median to median, and boundary curve to boundary curve.

Our new 3-manifold, which we again call M, has the same boundary as the old one did. The major difference is that the old median C, which is freely homotopic to several simple loops on ∂M, is now the $q^{\underline{th}}$ power of C'.

2. *Fuchsian groups*

In this section we construct the Kleinian groups whose 3-manifolds are the I-bundles of type (i).

2.1. A Fuchsian group acting on the upper half plane $\mathbf{H}^2$ is a discrete group of fractional linear transformations $z \longmapsto (az+b)/(cz+d)$, $ad-bc \neq 0$, where a,b,c,d are all real. The group is of the first kind if every real point is a limit point of the group; it is of the second kind otherwise.

2.2. Every Möbius transformation can be written as a composition of reflections and inversions in circles. Each of these elementary motions can be extended to act on the upper half space $\mathbb{H}^3 = \{(z,t) | z \in \mathbb{C},\ t \in \mathbb{R},\ t > 0\}$ as follows. If r is reflection in the line L, then the extension $r' : \mathbb{H}^3 \longrightarrow \mathbb{H}^3$ is given by $r'(z,t) = (r(z),t)$; if r is inversion in the circle C, then the extension $r' : \mathbb{H}^3 \longrightarrow \mathbb{H}^3$ is inversion in the 2-sphere which is orthogonal to $\mathbb{C}$ and which passes through C.

From here on we will regard groups of Möbius transformations as acting on both the extended complex plane $\hat{\mathbb{C}}$, and on hyperbolic 3-space $\mathbb{H}^3$.

2.3. A Möbius transformation which preserves $\mathbb{H}^2$ can be written as a product of reflections in lines orthogonal to the real line and inversions in circles centered at real points. It is easy to see that these motions preserve each Euclidean plane passing through the real line. It follows that if G is a Fuchsian group, then $\mathbb{H}^3/G = \mathbb{H}^2/G \times (0,1)$.

It is classical that if S^0 is the interior of a compact orientable surface with boundary, then there is a Fuchsian group G with $\mathbb{H}^2/G = S^0$. In fact, if we adjoin those real points where G acts discontinuously (the set of all points where G acts discontinuously is denoted by $\Omega = \Omega(G)$), we obtain $\mathbb{H}^2 \cup (\Omega \cap \hat{\mathbb{R}})/G = S$, the compact surface whose interior is S^0 (we use the notation $\hat{\mathbb{R}} = \mathbb{R} \cup \{\infty\}$).

2.4. A group of Möbius transformations is *elementary* if it has at most two limit points; a hyperbolic cyclic group H has two limit points and $\mathbb{H}^2/H$ is an annulus; the trivial group $\{1\}$ has no limit points and $\mathbb{H}^2/\{1\}$ is a disc.

Let S be the interior of a compact orientable surface with boundary, S neither a disc nor an annulus. Then there is a purely hyperbolic non-elementary Fuchsian group of the second kind, G, so that $\mathbb{H}^3/G = S \times I$. Conversely, if G is a finitely generated, non-elementary, purely hyperbolic Fuchsian group of the second kind, then $\mathbb{H}^2/G$ is the interior S of a compact orientable surface with boundary, neither a disc nor an annulus, and so $\mathbb{H}^3/G = S \times I$.

We can easily construct a group G as above. Let p be the genus of S and n the number of boundary components. We start with $4p + 2n - 2$ disjoint circles

$C_1, C'_1, \ldots, C_{2p+n-1}, C'_{2p+n-1}$, which bound a common region D, where each circle has

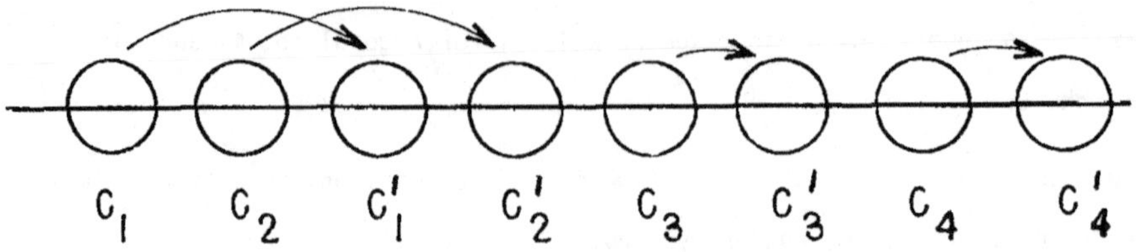

Figure 1

its center on the real axis, and all the circles have the same radius. Reading from left to right, we label these circles as $C_1, C_2, \ldots, C_{2p}, C'_1, C'_2, \ldots, C'_{2p}, C_{2p+1}, C'_{2p+1}, \ldots, C_{2p+n-1}, C'_{2p+n-1}$. (See Figure 1 for $p = 1$, $n = 3$.) We find a Möbius transformation a_i identifying C_i with C'_i as follows: a_i is composed of inversion in C_i followed by reflection in the perpendicular bisector of the centers of C_i, C'_i. Using either combination theorems [3,4] or Poincaré's polygon theorem [6], one easily sees that the group G generated by $a_1, \ldots, a_{2p+n-1}$ is discrete, acts freely on $\mathbb{H}^2$, and $\mathbb{H}^2/G$ is topologically equivalent to S.

2.5. A Fuchsian group G of the second kind acts discontinuously on certain segments of the real axis. Each such segment L is kept invariant by a hyperbolic cyclic subgroup H of G, and every element of G not in H maps L onto some other segment. A generator h of H is called a *boundary element* of G; H is called a *boundary subgroup* of G.

In general, if G is a group of motions of a space X, a subset $A \subset X$ is said to be *precisely invariant* under the subgroup H in G if $h(A) = A$ for every $h \in H$ and $g(A) \cap A = \emptyset$ for every $g \in G$, $g \notin H$.

Returning to our Fuchsian group G, it is easy to see that if C is the axis of H (i.e., C is the non-Euclidean line joining the fixed points of H), then the non-Euclidean half plane A between C and L is precisely invariant under H in G.

If instead of the axis, we look at the arc C' of a circle passing through the fixed points of H and lying in $\mathbb{H}^2$, then the region A' between C' and L may no longer be precisely invariant under H in G. If C' lies between C and L, then of course it is, but if C' lies far enough on the other side of C, it won't be.

We define the *lens angle* φ_H to be the smallest angle between C' and the real axis, where the region A between C' and L is precisely invariant under H; note that $\varphi_H \leqslant \pi/2$.

We now fix p and n, look at the construction of the preceding section, but in the unit disc model of the hyperbolic plane. We keep the circles equidistant from each other and let their size increase until the circles almost touch. We observe that as we do this the axes of the boundary subgroups get (Euclideanly) smaller. Going back to $\mathbb{H}^2$ and normalizing so that some boundary subgroup H has fixed points at 0 and ∞, we see that $\varphi_H \to 0$ as the circles get closer; or equivalently, $\varphi_H \to 0$ as the multiplier of the boundary element $h \to 1$ (fractional linear transformations which are conjugate to $z \mapsto \lambda z$ are called loxodromic; the number λ is called the *multiplier* of the transformation. If λ is real and positive, the transformation is sometimes called *hyperbolic*).

2.6. We actually need a somewhat stronger result than the preceding one.

The n-tuple of sets $(A_1, \ldots, A_n)$ is said to be *precisely invariant* under the n-tuple of subgroups $(H_1, \ldots, H_n)$ of G if the following hold.

(i) Each A_i is precisely invariant under H_i in G, and

(ii) for $i \neq j$ and for any $g \in G$, $g(A_i) \cap A_j = \emptyset$.

A Fuchsian group of the second kind G is called *isoholic* if all boundary elements of G have the same multiplier.

Now let G be an isoholic Fuchsian group of the second kind. We define the *lens angle* φ_G as the infimum of all angles φ with the following property. We pick a complete set of non-conjugate boundary subgroups $H_1, \ldots, H_n$. For each such boundary subgroup H_i with segment of discontinuity L_i, we construct the circular arc C_i lying in $\mathbb{H}^2$, having its end points at the fixed points of H_i, and making an angle of $\pi-\varphi$ with L_i. Let A_i be the region between L_i and C_i. We require that $(A_1, \ldots, A_n)$ be precisely invariant under $(H_1, \ldots, H_n)$ in G.

If we fix the topological type (p,n) of G, then it is easy to see that we can choose a path of isoholic Fuchsian groups of this topological type with the following two properties. *The multiplier* λ *of the boundary elements tends to* 1, *and the lens angle* $\varphi_G \longrightarrow 0$. To see this, we construct G as acting on the unit disc, and we choose some standard fundamental polygon for G (see for example Keen [2]). We choose our deformed groups so that $\lambda \longrightarrow 1$, the groups are all isoholic, have topologically equivalent fundamental polygons, and the Euclidean radii of the circles forming the boundary of their polygons stay bounded away from zero (this result also follows from Matelski's theorem [9]).

3. *Extended Fuchsian Groups*

In this section, we construct the Kleinian groups corresponding to the I-bundles of type (ii).

3.1. A non-elementary Kleinian group which is not itself Fuchsian, but contains a subgroup of index 2 which is Fuchsian, is called an *extended Fuchsian group*. If we start with a finitely generated, non-elementary, purely loxodromic extended Fuchsian group G then it is almost obvious that $\mathbb{H}^3/G$ is an I-bundle of type (ii). Of course we need the additional hypothesis that G^0, the Fuchsian subgroup of index 2, is of the second kind. We pick some element $g \in G$, $g \notin G^0$, and let j be the Möbius transformation of order 2 whose fixed points agree with the fixed points of g. Then $j \circ g$ preserves $\mathbb{H}^2$. By renormalizing so that j and g have fixed points at 0 and ∞, we see that g maps a Euclidean plane passing through the real line and inclined at an angle α to the upper half plane onto a Euclidean plane also passing through the real line and inclined at the same angle α to the lower half plane. The plane orthogonal to the real axis is kept invariant, and since G is torsion-free, the action of g on this plane can have fixed points only on the real axis. We conclude that $\mathbb{H}^3/G$, which is just $\mathbb{H}^3/G^0$ modulo the action of g, is an I-bundle of type (ii) over $\mathbb{H}^2/G^0$.

3.2. The converse statement is also true but is somewhat more difficult; that is, if P is an I-bundle of type (ii), then there is an extended Fuchsian group G so that $\mathbb{H}^3/G$ is homeomorphic to the interior of P.

We can assume that a conformal structure has been chosen on S so that h, the orientation reversing involution of S, is anti-conformal. We choose a Fuchsian group G^0 so that $\mathbb{H}^2/G^0 = S$, and we find an orientation reversing isometry j of $\mathbb{H}^2$ which projects to h on S. Let r be the reflection $z \longrightarrow \bar{z}$. We easily see that G, the group generated by G^0 and $r \circ j$ has the desired properties; i.e., $\mathbb{H}^3/G$ is homeomorphic to $P = S \times I/h'$, where $h'(x,t) = (h(x), 1-t)$.

3.3. Extended Fuchsian groups also have boundary subgroups and boundary elements. In fact, H is a boundary subgroup of the extended Fuchsian group G if an only if H is a boundary subgroup of G^0, the index 2 Fuchsian subgroup of G.

The definition of the lens angle φ_G is essentially the same as in 2.6; the only difference is that here we look at the union of a set in $\mathbb{H}^2$ and its mirror image in $\bar{\mathbb{H}}^2$, the lower half plane. While it may not be true that $\varphi_G = \varphi_{G^0}$, it remains true that for bounded topology, we can find isoholic extended Fuchsian groups with boundary multiplier $\lambda \longrightarrow 1$, and lens angle $\varphi_G \longrightarrow 0$.

4. *Building blocks*

In this and subsequent sections we illustrate the use of combination theorems to construct the Kleinian group corresponding to the 3-manifold construction given in §1.

4.1. We will illustrate the construction with a simple example, and leave the general construction to §8. For our example we have one surface S, a torus with two holes, and two I-bundles over S. P_1 is $S \times I$, and P_2 is an I-bundle of type

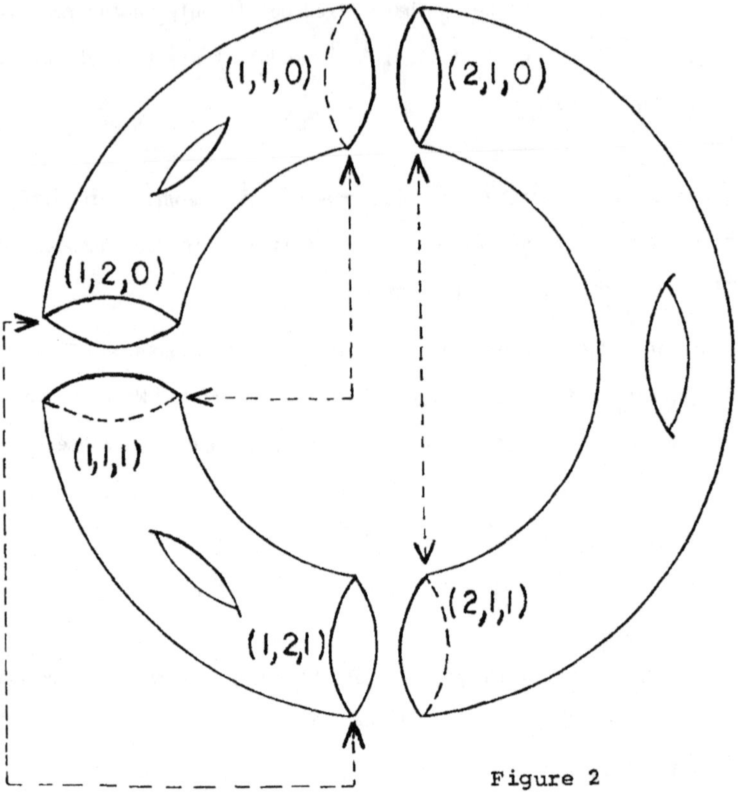

Figure 2

(ii) over S; since there is essentially only one such I-bundle, we need not specify P_2 further. Now $P_1 \cup P_2$ has six distinguished boundary curves which we label as (1,1,0), (1,2,0) on $S \times \{0\}$ in P_1, (1,1,1), (1,2,1) on $S \times \{1\}$ in P_1, and (2,1,0), (2,1,1) on P_2. The I-pairings are (1,1,0) I-paired with (1,1,1); (1,2,0) I-paired with (1,2,1); and (2,1,0) I-paired with (2,1,1) (here (i,j,k) is "the j-th boundary loop on P_i" $\times \{k\}$).

There are several possibilities for the S-pairing. The one we will use is (1,1,0) S-paired with (2,1,0); (1,2,0) S-paired with (1,1,1); and (1,2,1) S-paired with (2,1,1). When we glue together the favored boundaries using this pairing, we get a single closed surface of genus 4 (see Figure 2 where the S-pairings are shown by juxtaposition, and the I-pairings are shown by dotted curves).

4.2. We notice that with these pairings all six distinguished boundary curves fall into one cycle; i.e., our final 3-manifold has only one median. We do the first gluing, using combination theorem I in §5. The last two gluings, using the second combination theorem are in §6, and then the Dehn surgery type operation corresponding to the complex twist p/q is done in §7.

We choose a Fuchsian group G_1 representing P_1 (i.e., $\mathbb{H}^3/G_1 = \mathrm{int}(P_1)$), and we choose an extended Fuchsian group G_2 representing P_2. We choose G_1 and G_2 to be isoholic with the same multiplier λ; and we choose G_1 and G_2 so that they both have lens angle $\varphi < \pi/3$. (If we had $2n$ distinguished boundary loops in our cycle, we would choose $\varphi < \pi/n$.)

4.3. We need to draw pictures of G_1 and G_2. We start with G_1 and normalize

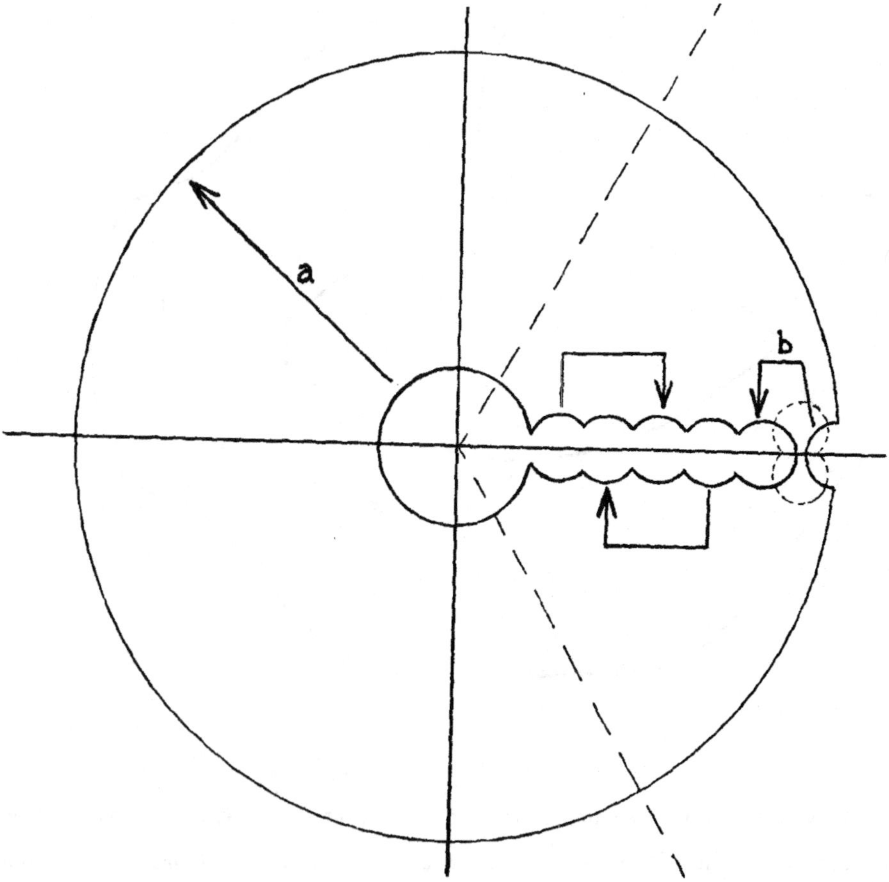

Figure 3

it so that some boundary element corresponding to the distinguished boundary curves (1,1,0) and (1,1,1) is of the form $z \longmapsto \lambda z$, $\lambda > 1$. As part of this normalization we also require that the negative reals be an interval of discontinuity.

We draw a fundamental domain D_1 for G_1. This is shown in Figure 3. The domain shown is a fundamental domain for G_1 as a Kleinian group; it is constructed by starting with a fundamental polygon for G_1 in $\mathbb{H}^2$, and then adjoining the reflected polygon in $\overline{\mathbb{H}}^2$. The dotted rays and circles illustrate the lens angle φ.

In Figure 3, we have distinguished two boundary elements a and b of G_1 for later use. The transformation a is $z \longmapsto \lambda z$, and b is the other boundary transformation which identifies the sides of D_1; there is no particular reason to choose this transformation as b rather than its inverse.

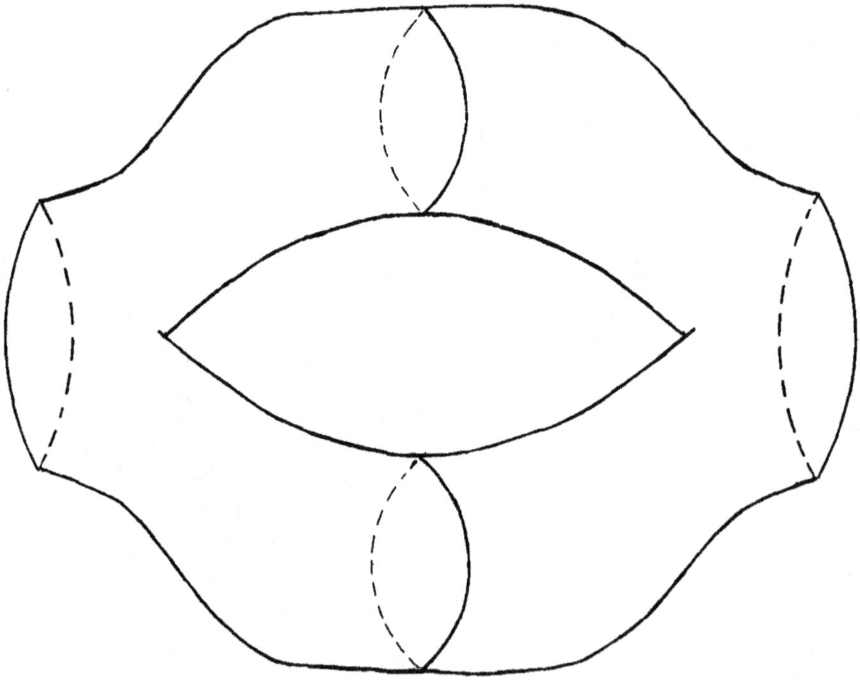

Figure 4

4.4. It is somewhat more complicated to get a picture of P_2. The easiest way is to start by cutting S in half along the two loops shown in Figure 4. This gives us two 3-holed spheres S_2' and S_2'', which we assume to be mirror images of each other. We start with a Fuchsian group G_2', for which $\mathbb{H}^2/G_2' = S_2'$. A fundamental domain is shown in Figure 5. The dotted circles represent the axes of two of the boundary elements of G_2'. We adjoin an element g_2 to G_2', where g_2 maps one of

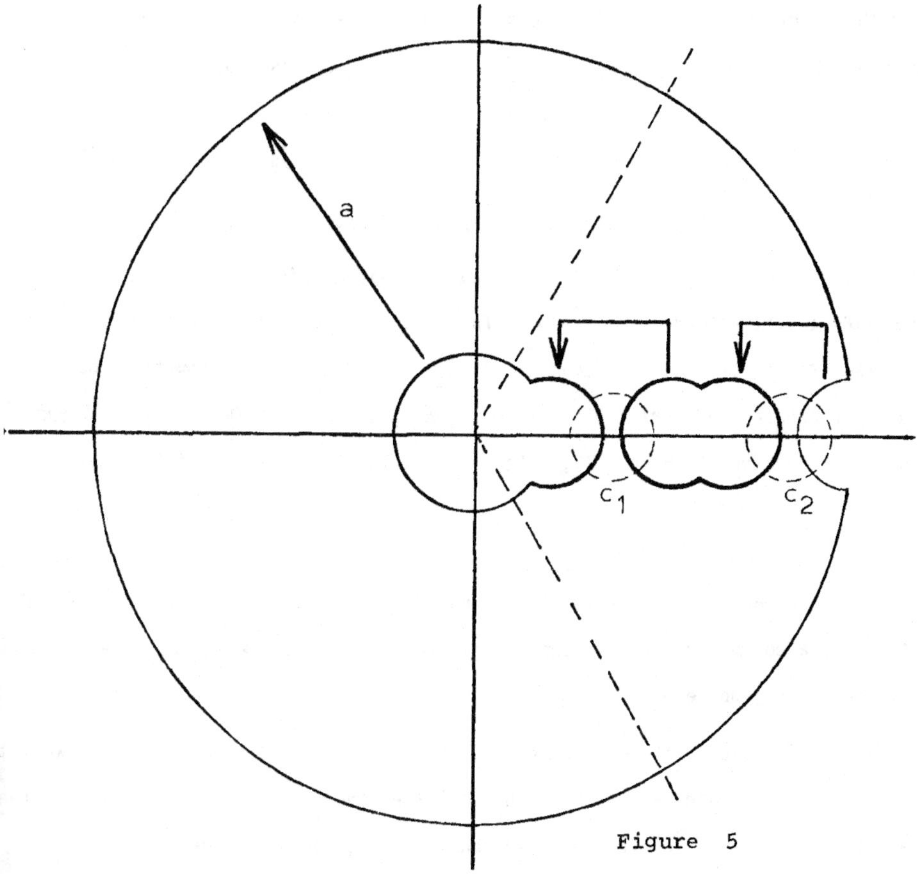

Figure 5

these axes onto the other, mapping the outside of one onto the inside of the other, and g_2 maps the upper half plane onto the lower half plane; i.e., call these axes C_1 and C_2 and assume they have the same Euclidean radius, then g_2 is the composition of first inversion in C_1, then euclidean translation taking C_1 onto C_2, then reflection in the real axis. It is not hard to see that G_2, the group generated by G_2' and g_2, is an extended Fuchsian group; $\mathbb{H}^3/G_2 = P_2$; the fundamental domain for G_2 is just the intersection of the fundamental domain for G_2' with the exterior of C_1 and C_2.

5. *First Combination Theorem*

In this section we use the first combination theorem to construct a Kleinian group corresponding to gluing P_1 to P_2, where $(1,1,0)$ is glued to $(2,1,0)$.

5.1. For our purposes, the first combination theorem says the following [3,5].

Let G_1, G_2 *be Kleinian groups with a common subgroup* H. *Suppose there is a simple closed curve* C *dividing* $\hat{\mathbb{C}}$ *into two topological discs* B_1, B_2 *where* B_i *is precisely invariant under* H *in* G_i. *Then* G, *the group generated by* G_1 *and* G_2 *is discrete,* G *is the free product of* G_1 *and* G_2 *with amalgamated subgroup* H, *and if we choose fundamental domains* D_i *for* G_i, *where* $D_i \cap B_i$ *is a fundamental domain for the action of* H *on* B_i, *then* $D = D_1 \cap D_2$ *is a fundamental domain for* G.

5.2. In our particular case we already have G_1 and G_2, which we now call G_2^*, and they have a common cyclic subgroup H, generated by $a : z \longmapsto \lambda z$, but they do not satisfy the other hypotheses.

We leave G_1 alone, and we change G_2^* by conjugation. Instead of G_2^*, we let $g : z \longmapsto \exp(2\pi i/3)z$, and we let $G_2 = gG_2^*g^{-1}$. Then the limit set of G_2 lies on the ray $\arg z = 2\pi/3$. We have already noted that a is a common boundary element of G_1 and G_2.

We let C be the circle formed from the rays $\arg z = \pi/3, 4\pi/3$. This circle separates $\hat{\mathbb{C}}$ into two discs: B_1 which contains the ray $\arg z = 2\pi/3$, and B_2 which contains the positive reals.

Since we chose our lens angle $\varphi < \pi/3$, B_1 is precisely invariant under H in G_1, and B_2 is precisely invariant under H in G_2. Hence, by combination theorem I, the group G generated by G_1 and G_2 is discrete. A fundamental domai for G is shown in Figure 6, where the rays $\arg z = 0$, $\pi/3$, $2\pi/3$, and $4\pi/3$ are dotted.

5.3. We remark that G is an extended quasifuchsian group of the second kind (i.e., a quasiconformal deformation of an extended Fuchsian group of the second kind).

The non-Euclidean plane L, whose boundary is C, divides $\mathbb{H}^3$ into two half spaces M_1 and M_2, where M_i is precisely invariant under H in G_i. We get a picture of $\mathbb{H}^3/G$ as follows. We delete M_1 and all its translates under G_1 from $\mathbb{H}^3$, call the result N_1. Then N_1/G_1 is a proper subset of P_1, and contains L/H on its boundary. Similarly, delete M_2 and all its translates under G_2 from $\mathbb{H}^3$ and call the remainder N_2. Then N_2/G_2 is a proper subset of P_2 which contains L/H on its boundary. The 3-manifold $\mathbb{H}^3/G$ is $N_1/G_1 \cup N_2/G_2$ with the boundaries L/H identified.

It's easy to see that N_1/G_1 is homeomorphic to P_1, and under this homeomorphism, L/H corresponds to the boundary cylinder $(1,1,0) \times I$. Similarly, N_2/G_2 is homeomorphic to P_2, and L/H corresponds to $(2,1,0) \times I$. The gluing along L/H glues $(1,1,0)$ to $(2,1,0)$ as we want, but also glues $(1,1,1)$ to $(2,1,1)$ which we don't want. We'll rectify this with the next operation.

6. *Second Combination Theorem*

In this section, we use the second combination theorem to complete the gluing of P_1 to P_2.

6.1. The version of Combination Theorem II [4,5] that we will use asserts the following.

Let G *be a Kleinian group and let* H_1 *and* H_2 *be subgroups of* G. *Let* B_1 *and* B_2 *be disjoint topological discs where the pair* (B_1,B_2) *is precisely invariant under the pair* (H_1,H_2) *in* G. *Suppose there is a Möbius transformation* f *mapping the interior of* B_1 *onto the exterior of* B_2, *where* $fH_1f^{-1} = H_2$. *Then* G^*, *the* *group generated by* G *and* f, *is discrete; the relations of* G^* *are those of* G *together with* $fH_1f^{-1} = H_2$. *A fundamental domain for* G^* *can be obtained as follows. Let* D *be a fundamental domain for* G, *where* $D \cap B_i$ *is a fundamental domain for the action of* H_i *on* B_i; *then* $D^* = D \cap \text{ext}(B_1) \cap \text{ext}(B_2)$ *is a fundamental domain for* G^*.

6.2. We start with the group G constructed in §5. The first subgroup H_1 is generated by $a : z \mapsto \lambda z$, which is the common subgroup of G_1 and G_2, and the

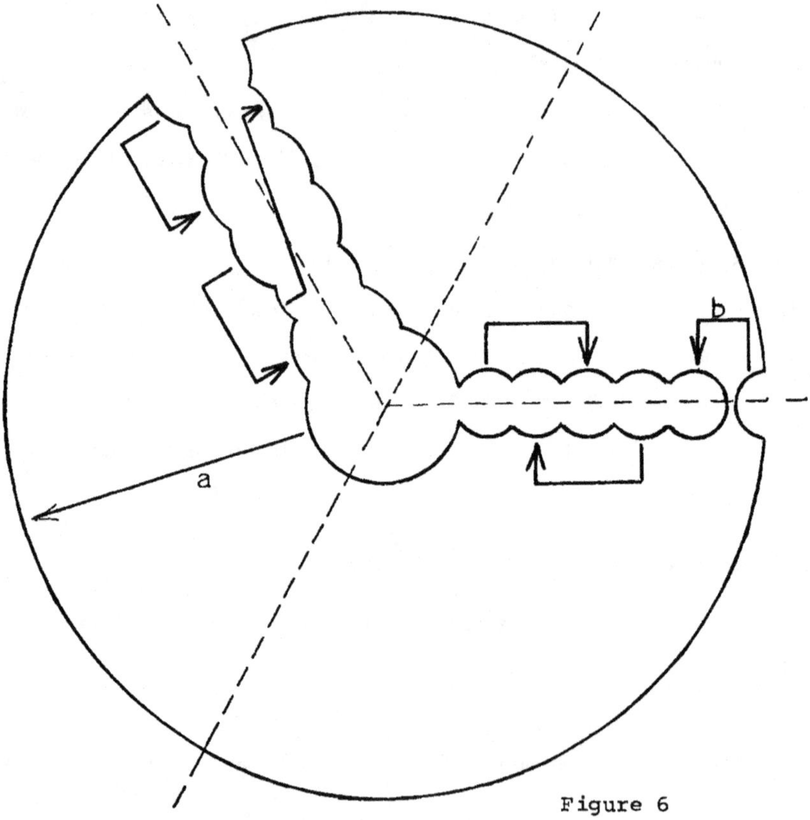

Figure 6

second subgroup H_2 is the boundary subgroup of G_1, generated by the transformation b shown in Figures 3 and 6. We have chosen a and b so that they have the same multiplier λ, and hence are conjugate in the group of all Möbius transformations.

We next use precise invariance and our lens angle $\varphi < \pi/3$ to set up the topological discs B_1 and B_2. We construct B_1 as the sector $|\arg z - \frac{4\pi}{3}| < \varphi$. Since we know that the entire sector $\pi < \arg z < \frac{5\pi}{3}$ is precisely invariant under H_1 in G, B_1 is precisely invariant under H_1 in G.

We construct B_2 by constructing an arc of a circle C with end points at the fixed points of b; C lies in $\mathbb{H}^2$; C makes an angle of $\pi - \varphi$ with L, the segment of discontinuity of H_2. Then B_2 is the region between C and L,

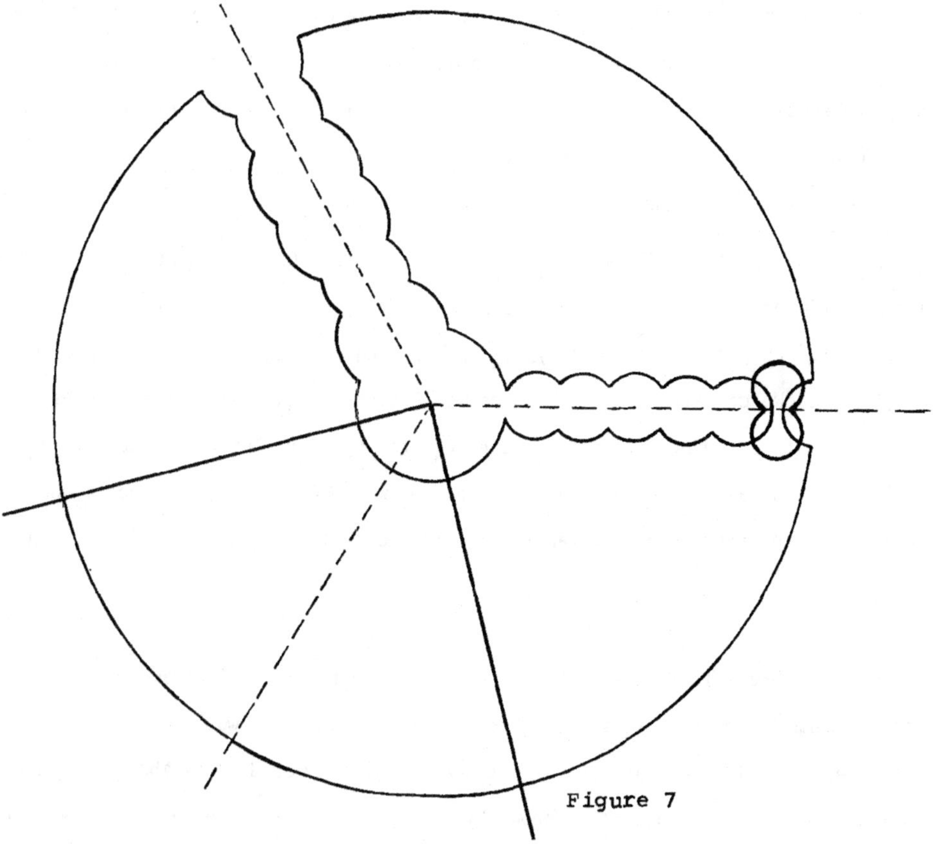

Figure 7

together with its mirror image in the lower half plane. In Figure 7, B_1 and B_2 are bounded by heavy lines and circles; the dotted lines are $\arg z = 0$, $2\pi/3$, $4\pi/3$; the rest of Figure 7 is the same as Figure 6, except that the identifications of the sides are not shown.

6.3. We let x be the attractive fixed point of b. It is immediate that there are two distinct 1-parameter families of Möbius transformations which map $\text{int}(B_1)$ onto $\text{ext}(B_2)$ and which conjugate H_1 to H_2. The elements of the first family map 0 to x, and map $\arg z = \frac{4\pi}{3} - \varphi$ into the upper half plane, while the elements of the other family map ∞ to x and take the ray $\arg z = \frac{4\pi}{3} - \varphi$ into the lower half plane.

The combination theorem is insensitive to this choice. All elements in both families work equally well. Topologically, there are only two choices, all choices from the same 1-parameter family give topologically equivalent results. To see which choice we should make, we need to go back to the S-pairing pattern that we chose in 4.1. To do this we look at the projections of the rays $C_1 = \{\arg z = \pi/3\}$, $C_2 = \{\arg z = \pi\}$, $C_3 = \{\arg z = 5\pi/3\}$, and the arcs $C_4 = \partial B_2 \cap \mathbb{H}^2$, $C_5 = \partial B_2 \cap \overline{\mathbb{H}}^2$.

If we look just at G_1 and P_1, then we can view C_1, C_3, C_4, C_5 as projecting to distinguished boundary curves on the boundary of P_1, with C_1 projecting to $(1,1,0)$, C_3 projecting to $(1,1,1)$, C_4 projecting to $(1,2,0)$, and C_5 projecting to $(1,2,1)$. We can also view our two possibilities for the use of the combination theorem as either gluing C_2 to C_4 and C_3 to C_5, or gluing C_2 to C_5 and C_3 to C_4. The S-pairing we chose has C_3 paired with C_4; hence we should choose an element f from the second 1-parameter family, so that f glues C_2 to C_5 and C_3 to C_4.

6.4. We remark that if we chose an element from the first 1-parameter family, then the 3-manifold M corresponding to the final group G^* would again be composed of the two I-bundles P_1 and P_2, but now the gluing would be as shown in Figure 8 (as in Figure 2, the S-pairing is shown by juxtaposition, and the I-pairing by dotted curves). Note that in this case, ∂M has two components, one a surface of genus 3 and one a surface of genus 2.

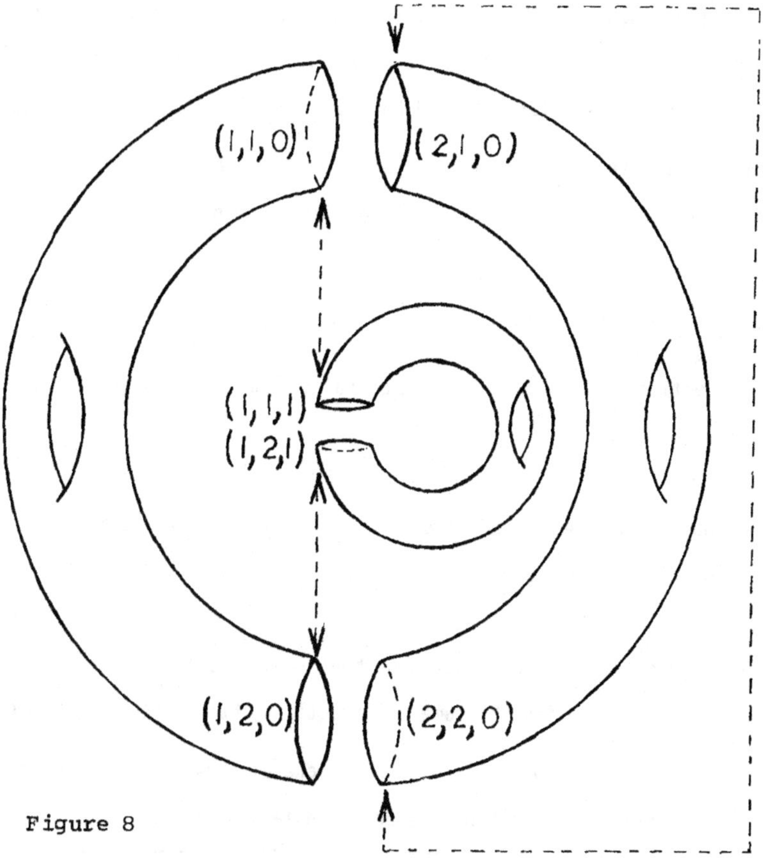

Figure 8

6.5. The regions B_1 and B_2 are not circular discs, but there are obvious regions in $\mathbb{H}^3$ corresponding to them. Each B_i is bounded by two circular arcs; let M_i be the region in $\mathbb{H}^3$ bounded by the non-Euclidean planes passing through these arcs, where B_i is the (Euclidean) boundary of M_i. It's easy to see that (M_1, M_2) is precisely invariant under (H_1, H_2) in G.

We remove $M_1 \cup M_2$ and all its translates under G from $\mathbb{H}^3$, call the remainder N. Then N is invariant under G and N/G is topologically equivalent to $\mathbb{H}^3/G$. Then $\mathbb{H}^3/G^*$ is just N/G with $\partial M_1/H_1$ identified with $\partial M_2/H_2$.

At the end of 5.3 we observed that $\mathbb{H}^3/G$ was $P_1 \cup P_2$ with the distinguished boundary curves (1,1,0) glued to (2,1,0), as it should be, but also with (1,1,1) glued to $(2,1,1)$. Removing M_1 breaks this last gluing, introduces a cylinder $\partial M_1/H_1$ into the boundary of our 3-manifold, where the boundary of the cylinder can be identified with the two distinguished boundary curves $(1,1,1)$ and $(2,1,1)$.

The other cylinder $\partial M_2/H_2$ can be identified with the cylinder $(1,2,0) \times I$ on ∂P_1. Gluing these together gives us the 3-manifold we want.

7. *Complex twists*

In this section, we perform the last operation corresponding to the complex twist.

7.1. We now assume that in addition to P_1, P_2, and the S-pairing, we are also given the complex twist $p/q \neq 0$.

7.2. The group G^* was constructed without regard to the complex twist; we need to construct a new G^*, using the same constructions, but different positions.

We first construct a new G_1 and G_2^*, where all boundary elements of G_1 and G_2^* have the same multiplier, but now we need the lens angle $\varphi < \pi/3q$.

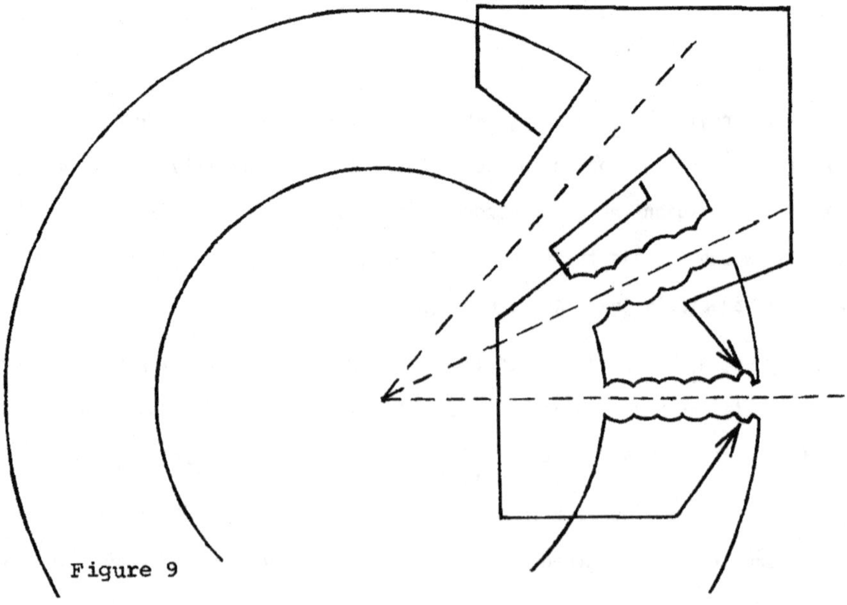

Figure 9

We keep the new G_1 normalized exactly as the old one was. Then G_2 is G_2^* conjugated by $z \longmapsto \exp(2\pi i/3q)z$. Exactly as before we can use combination theorem I to form G, the group generated by G_1 and G_2. We then use combination theorem II, with H_1 and H_2 as before, but now B_1 is the sector $|\arg z - 4\pi/3q| < \varphi$, and B_2 is as before but with the new lens angle φ. A fundamental domain for the new G^* is shown in Figure 9, where we have chosen $p/q = 2/5$ (in Figure 9, we show only the identifications not previously displayed). We remark that this new group G^* is topologically indistinguishable from the old G^*.

7.3. We next observe that the sector $5\pi/3q < \arg z < 2\pi - \pi/3q$ is precisely invariant under H_1, the subgroup of G^* generated by $a : z \longmapsto \lambda z$.

We let H_0 be the group generated by $a_0 : z \longmapsto \lambda^{1/q} \exp(2\pi ip/q)z$, where $\lambda^{1/q} > 0$. We observe that $a_0^q = a$, and so H_1 is a subgroup of H_0 of index q. In particular G^* and H_0 have H_1 as a common subgroup.

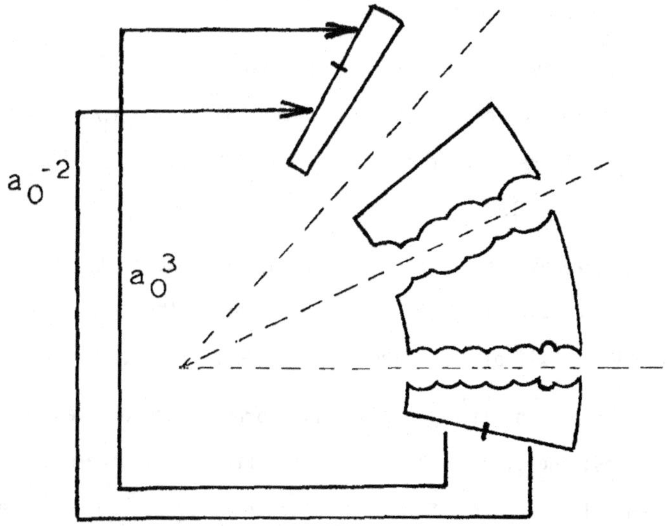

Figure 10

We let C_1' be the ray $\arg z = 5\pi/3q$ and we let C_2' be the ray $\arg z = -\pi/3q$. We have already observed that the region B_1', lying between C_1' and C_2' and containing the negative reals is precisely invariant under H_1 in G^*.

We let B_2' be the other region bounded by C_1' and C_2'. Since H_0 and H_1

are cyclic, it is immediate that B_2' will be precisely invariant under H_1 in H_0 if and only if the sector $B_2'' = \{0 < \arg z < 5\pi/3q + \pi/3q = 2\pi/q\}$ is. Of course p and q are relatively prime, and so $a_0^m(z) = \lambda^{m/q}\exp(2\pi imp/q)z$ cannot map B_2'' back to itself unless $m \equiv 0 \pmod q$.

We conclude that B_2' is precisely invariant under H_1 in H_0. Hence we can form the group $\tilde{G}_1$ generated by G^* and H_0 via combination theorem I. In Figure 10, we show a fundamental domain for $\tilde{G}$; the identifications which are given in Figure 9 are not shown in Figure 10. Note that the two new sides each have a vertex on them, and the segments are identified by different powers of a_0.

7.4. We next observe that in terms of 3-manifolds this construction is essentially the same as that given in 1.8. We define the solid sector in $\mathbb{H}^3$, $M_1 = \{(z,t) \mid z \in B_1', t > 0\}$, and let M_2 be the complementary sector.

We observe that M_1 is precisely invariant under H_1 in G^*, and that M_2 is precisely invariant under H_1 in H_0.

If we remove M_1 and all its translates under G^* from $\mathbb{H}^3$, then we are left with a set N_1 which is invariant under G^*, and N_1/G^* is homeomorphic to $\mathbb{H}^3/G^*$. We have simply replaced the cylinder of $\partial M_1/H_1$ on $\partial\mathbb{H}^3/G$ with the cylinder of $\partial N_1/H_1$ on $\partial N_1/G$.

If we remove M_2 and all its translates under H_0 from $\mathbb{H}^3$, we will be left with a 2-dimensional object N_2 consisting of the axis $\{z=0, t>0\}$ of H_0, together with the q half planes $\{\arg z = 2\pi r/q - \pi/q, t > 0\}$, $r = 1,\dots,q$.

The manifold $\mathbb{H}^3/H_0$ is a solid torus. Each of these half planes has two boundaries; one lying on the axis which projects to the median of the solid torus, and the other lying in $\mathbb{C}$. Since H_0 is generated by $z \longmapsto \lambda^{1/q}\exp(2\pi ip/q)z$, one easily sees that a ray from the origin projects to a loop on the torus which is homotopic to X^qY^p, where X,Y form a basis for the fundamental group of the torus, and Y is trivial in the solid torus.

Our use of the combination theorem corresponds to gluing N_1 to N_2, where the attaching map is given by the identification of $\partial M_1/H_1$ with $\partial M_2/H_2$.

In the construction above, we did not need N_2 to be 2-dimensional. We could have chosen M_2 to be a slightly smaller sector, N_2 would then be a 3-manifold with boundary. Of course this boundary is still a cylinder, and the new N_2/H_0 is homeomorphic to its boundary $\times$ I.

7.5. We remark that our original G^* had three components kept invariant by the trnasformation a (one component contained in $0 < \arg z < 2\pi/3q$, another in $2\pi/3q < \arg z < 4\pi/3q$, and the third in $4\pi/3q < \arg z < 2\pi$). Our final group $\tilde{G}$ has $3q$ components kept invariant by the transformation a; these are contained in sectors of the form $2\pi r/3q < \arg z < 2\pi(r+1)/3q$. The fixed points 0 and ∞ of a_0 lie on the boundary of all these components, but a_0 keeps no component invariant; the number of components of $\tilde{G}$ modulo the action of H_0 is precisely three.

8. *General construction*

In the preceding sections, we constructed a particular Kleinian group corresponding to a particular collection of I-bundles with S-paired distinguished boundary curves, and given complex twists. In this section we discuss the general construction.

8.1. We now suppose that we are given a finite collection of I-bundles $P_1,\dots,P_n$ with S-paired distinguished boundary curves satisfying the restrictions of 1.4. By alternately following the I-pairing and the S-pairing, the distinguished boundary curves fall into cycles; we assume that for each cycle, we are given a complex twist p/q, $0 \leqslant p/q < 1$ ($p/q = 0$ means that there is no complex twist associated to the cycle; i.e., $q = 1$).

8.2. We choose isoholic Fuchsian and extended Fuchsian groups $G_1,\dots,G_n$, corresponding to $P_1,\dots,P_n$, where all boundary elements have the same multiplier λ, and all these groups have the same lens angle φ. We choose this lens angle φ as follows. Any given cycle has $2m$ distinguished boundary curves in it, and complex twist p/q; we require $\varphi < \pi/mq$.

8.3. We pick some distinguished boundary curve W_1', let W_2 be S-paired with W_1'; let W_2' be I-paired with W_2; let W_3 be S-paired with W_2', etc. We temporarily assume that $W_2' \neq W_1$, the distinguished boundary loop I-paired with W_1'.

We assume that W_1 lies on P_1; we normalize G_1 so that $z \longmapsto \lambda z$ is a boundary element corresponding to W_1, and so that the negative reals is an interval of discontinuity for G_1.

If W_2 does not lie on P_1, then assume it lies on P_2, we conjugate G_2 so that its limit set lies on the ray $\arg z = 2\pi/mq$, and so that the boundary element corresponding to W_2 is $z \longmapsto \lambda z$. We then apply combination theorem I.

If W_2 lies on P_1, we use combination theorem II with H_1 generated by $z \longmapsto \lambda z$, H_2 a boundary subgroup of G_1 corresponding to W_2. We choose B_1 to be the sector $\{|\arg z - 2\pi/mq| < \varphi\}$, and B_2 to be the corresponding region precisely invariant under H_2.

We continue as above. If W_3 lies on say P_3, where the corresponding G_3 has not as yet been used, then we normalize G_3 so that its limit set lies on $\arg z = 4\pi/mq$ and use combination theorem I. If W_3 lies on P_1 or P_2 which has already been used, then we use combination theorem II, with the sector $\{|\arg z - 4\pi/mq| < \varphi\}$ being mapped onto the exterior of an appropriate sector, between two arcs of circles, precisely invariant under a boundary subgroup in G_1 or G_2.

We continue as above until we have used the combination theorems m-1 times. At that point we note that the sector $\{(2m-1)\pi/mq < \arg z < (2mq-1)\pi/mq\}$ is precisely invariant under the group generated by $z \longmapsto \lambda z$ in the resulting group, and so we can use combination theorem I to adjoin the appropriate q-th root of $z \longmapsto \lambda z$; i.e., $z \mapsto \lambda^{1/q} \exp(2\pi i p/q) z$.

8.4. Our collection of I-bundles might have more than one cycle. We need to repeat the above procedure for each cycle. We remark that for the second and subsequent cycles, the initial normalization takes some care; in particular, we could conjugate G_1 by $z \longmapsto 1/z$.

We let $\tilde{G}_1$ be the final group constructed as above for the first cycle. We assume that the I-bundle P_1, corresponding to $G_1 \subset \tilde{G}_1$ has a distinguished boundary curve W on it, where W does not lie in the first cycle. As before, we normalize $\tilde{G}$ so that G_1 has all its limit points on the positive reals (including 0 and ∞), and so that $a : z \longmapsto \lambda z$ corresponds to W.

In order to start the construction for the second cycle, we need to decide if $\arg z = \varphi$ projects to W or to W', I-paired with W. We do this by continuity from the first cycle, and then proceed with the construction as above.

8.5. We need to consider the possibility that in our cycle $m = 1$; i.e., the cycle only contains the two distinguished boundary curves W_1 and W_2, which are both I-paired and S-paired with each other.

If the complex twist $p/q = 0$, then our algorithm says that we normalize the Fuchsian or extended Fuchsian group corresponding to the I-bundle containing W_1 and W_2, and do nothing further with this cycle. Regardless of what we do with the other cycles, the result will always be a Kleinian group where one of the components of the set of discontinuity is not simply connected. This is the only case where this occurs; i.e., if we do not have such a cycle, then every component of the final group is simply connected.

8.6. There are two other special cases which should be mentioned. If there is a cycle with $m = 1$, $p/q = 1/2$, then we start with a Fuchsian or extended Fuchsian group G_1, and adjoin a square root of a boundary element, where the square root interchanges upper and lower half planes; i.e., we end up with an extended Fuchsian group G. Hence this operation is redundant; we could have started with $\mathbb{H}^3/G$ rather than $\mathbb{H}^3/G_1$.

Similarly if $m = 2$, $p/q = 0$; after performing the one operation, we end up with a quasi-fuchsian or extended quasi-fuchsian group; we could equally well have started with a topologically equivalent Fuchsian or extended Fuchsian group.

9. *Panelled groups*

In the preceding sections we constructed certain Kleinian groups. In this section we give a characterization of this family of groups.

9.1. For the purpose of this paper, a Kleinian group is a finitely generated discrete group of Möbius transformations which acts discontinuously somewhere in $\hat{\mathbb{C}}$. The full set of discontinuity of the Kleinian group G is denoted by $\Omega = \Omega(G)$.

We are interested here only in Kleinian groups G which act freely in $\mathbb{H}^3$, and for which $M(G) = \mathbb{H}^3/G$ is the interior of a compact 3-manifold with boundary, where every component of the boundary has genus ≥ 2. The first statement means that G has no elliptic elements, and the second that G is non-elementary and contains no parabolic elements. From here on, we will consider only Kleinian groups which are purely loxodromic (including hyperbolic), and non-elementary.

9.2. The connected components of $\Omega(G)$ are called components of G. If every component of G is simply connected then G is called a *web group*. (This definition is incorrect if G has only one component (see Abikoff [1]); it is also incorrect if G is not both finitely generated and purely loxodromic.)

We have already remarked that except for the possibility of a cycle with $m = 1$, $p/q = 0$, the Kleinian groups in our class are all web groups.

9.3. Let G be a web group, let Δ be a component of G, and let H be the stabilizer of Δ in G; i.e., $H = \{g \in G | g(\Delta) = \Delta\}$. Since Δ is simply connected, there is a Riemann map $f : \Delta \longrightarrow \mathbb{H}^2$; we let $F = fHf^{-1}$; F is called the Fuchsian model of H.

Using the map f, and the Fuchsian model F, we can assign an axis A to every element of H, and we can construct the Nielsen convex region $K(J)$ for every subgroup $J \subset H$; $K(J)$ is the convex closure of the limit set of J (the convex closure of the limit set of fJf^{-1} is well defined; $K(J)$ is f^{-1} of that set).

We now look also at some other component Δ^* of G, with stability subgroup

H^*. We let $J = H \cap H^*$. It was shown in [7] that $K(J)$ is precisely invariant under J in H. Of course we also could look at $K^*(J)$, the Nielsen convex region of J in H^*, and get the same result; i.e., $K^*(J)$ is precisely invariant under J in H^*.

We can describe the result above as the following. *Either* $K(J)/J = \emptyset$, *or* $K(J)/J$ *is a geodesic on* Δ/H, *or* $K(J)/J$ *is a subsurface of* Δ/H *bounded by geodesics.*

9.4. Using the techniques of [7], one can generalize the above. Let Δ' be a third component and let H' be the stabilizer of Δ' in G. Let $J' = H \cap H'$, and let $K(J')$ be the Nielsen convex region of J' in Δ. There are only three possibilities. Either $K(J) = K(J')$, or $K(J) \cap K(J') = \emptyset$, or $J \cap J'$ is cyclic and $K(J) \cap K(J')$ is the axis of $J \cap J'$.

When we project to Δ/H, we get the following statement. Either $K(J)/J = K(J')/J'$, or $K(J)/J$ and $K(J')/J'$ have disjoint interiors; they may have some number of boundary geodesics in common.

9.5. Continuing with our assumption that G is a web group, we look at all the subsurfaces of the form $K(J)/J$ as $J = H \cap H^*$ varies over all pairs of component subgroups H, H^*. We say that G is *panelled* if the closure of the union of these subsurfaces (there are only finitely many distinct such subsurfaces) contains all of $\Omega(G)/G$.

9.6. We remarked in 8.5, that if in our 3-manifold, we had no cycles with $m = 1$, $p/q = 0$, then the corresponding Kleinian group G is a web group. It is not hard to prove in this case that G is panelled.

The converse is also true.

If G *is a (purely loxodromic) panelled web group, then* G *is a quasiconformal deformation of one of the groups described in* §8, *and* $\mathbb{H}^3/G$ *is one of the 3-manifolds described in* §1.

9.7. There is a more general notion of panelled group which applies to Kleinian groups with elliptic and parabolic elements, and to groups which are not necessarily web groups. The definition, constructions, and classification are closely related to the material in this paper. These more general results will appear elsewhere.

As part of the above program, we also obtain a set of invariants for the class of panelled web groups modulo quasi-conformal deformation. These invariants are precisely those given in §1 (with the restrictions of §8.5-6); i.e., the topological types of I-bundles, the I-pairing and S-pairing of the distinguished boundary loops, and the complex twists.

STATE UNIVERSITY OF NEW YORK
STONY BROOK, NY 11794

REFERENCES

[1] W. Abikoff, The residual limit sets of Kleinian groups, *Acta Math.* 130 (1973), 127-144, MR 53 #8413.

[2] L. Keen, Intrinsic moduli on Riemann surfaces, *Ann. of Math.* 84 (1966), 404-420.

[3] B. Maskit, On Klein's combination theorem, *Trans. Amer. Math. Soc.* 120 (1965), 499-509, MR 33 #274.

[4] ________, On Klein's combination theorem, II, *Trans. Amer. Math. Soc.* 131 (1968) 32-39, MR 36 #6618.

[5] ________, On Klein's combination theorem, III, *Advances in the theory of Riemann surfaces* (Proc. Conf. Stony Brook, N.Y., 1969), pp. 297-316. *Ann. of Math. Studies* No. 66, Princeton Univ. Press, Princeton, N.J., 1971, MR 44 #6955.

[6] ________, On Poincaré's theorem for fundamental polygons, *Advances in Math.* 7 (1971), 219-230, MR 45 #7049.

[7] ________, Intersections of component subgroups of Kleinian groups, *Discontinuous groups and Riemann surfaces* (Proc. Conf. Univ. Maryland, College Park, Md., 1973) pp. 349-367. *Ann. of Math. Studies*, No. 79, Princeton Univ. Press, Princeton, N.J., 1974, MR 50 #7514.

[8] ________, On the classification of Kleinian groups II: Signatures, *Acta Math.* 138 (1977), 17-42.

[9] J.P. Matelski, A compactness theorem for Fuchsian groups of the second kind, *Duke Math. J.* 43 (1976), 829-840.

PRESCRIBED MONODROMY ON NONCOMPACT SURFACES

R. MICHAEL PORTER

Consider a Fuchsian group G acting on the unit disk Δ. An orientation-preserving local homeomorphism $F : \Delta \to \hat{\mathbb{C}} = \mathbb{C} \cup \{\infty\}$ is a *topological deformation* of G if it semiconjugates G onto another subgroup of the Möbius group $\mathrm{Aut}\, \hat{\mathbb{C}}$, in other words, if there exists a (necessarily unique) group homomorphism $\chi : G \to \mathrm{Aut}\, \hat{\mathbb{C}}$ such that for each $A \in G$,

$$f \circ A = \chi(A) \circ f \tag{1}$$

identically in Δ. We will say that χ is induced by f.

Holomorphic deformations have been studied extensively ([2], [3], [4]) as they represent the projective structures on the quotient Riemann surface Δ/G. It is known that when Δ/G is compact, certain homomorphisms χ cannot be induced by holomorphic deformations (for example, when $\chi(G)$ is an affine subgroup of $\mathrm{Aut}\, \hat{\mathbb{C}}$); the reasoning used in the Corollary below then shows that they cannot be induced by topological deformations. We will show that when Δ/G is not compact, there are no topological obstructions for a homomorphism to be induced by a deformation.

THEOREM. *Let* G *be a Fuchsian group whose quotient space* Δ/G *is not compact. Let* $\chi : G \to \mathrm{Aut}\, \hat{\mathbb{C}}$ *be a group homomorphism. Then a necessary and sufficient condition that there exist a topological deformation of* G *which induces* χ *is that* $\mathrm{tr}\, \chi(E) = \mathrm{tr}(E)$ *for every elliptic element* E *of* G.

COROLLARY. *Let* S *be a noncompact topological surface and take any group homomorphism* $\chi : \pi_1(S) \to \mathrm{Aut}\, \hat{\mathbb{C}}$. *Then there exist a Fuchsian group* G *acting on* Δ *and a holomorphic deformation* $f : \Delta \to \hat{\mathbb{C}}$ *of* G *such that* Δ/G *is homeomorphic to* S *and* (1) *is satisfied when* $\pi_1(S)$ *is identified with* G *in the natural way.*

Proof. Choose a Fuchsian group G_0 acting on Δ such that Δ/G_0 is homeomorphic to S. Since G_0 has no elliptic elements, the Theorem gives us a topological deformation f_0 of G_0 inducing χ. The values of f define a conformal structure on S (not necessarily conformally equivalent to Δ/G_0), and by the Uniformization Theorem there exists a Fuchsian group G such that Δ/G is conformally equivalent to S. Since $\Delta \to \Delta/G$ and $\Delta \to \Delta/G_0$ are universal coverings, the conformal equivalence lifts to a topological map $h : \Delta \to \Delta$ conjugating G into G_0. It follows immediately that $f = f_0 \circ h$ is a holomorphic deformation of G which induces χ.

1. Before proving the Theorem we will fix some terminology and recall some well known properties of Fuchsian groups (see for instance [1]). We assume G is not the trivial group.

1.1. G has a simply connected fundamental domain in Δ whose relative boundary in Δ is a union of *edges*, which are arcs of circles orthogonal to the boundary of Δ. This domain together with its boundary will be called D. For each edge e there is a transformation $A_e \in G$ such that $e' = A_e(e)$ is another edge of D; the elements A_e generate G. Each edge is oriented by the requirement that D lie to its left, and as a mapping from e to e', A_e reverses orientation. An endpoint v of an edge will be called a vertex when it lies in Δ. Thus each vertex belongs to two edges whereas an edge may have 0, 1, or 2 vertices as endpoints.

The vertices form equivalence classes called *cycles*, which may be described as follows. Beginning with a vertex v_1 we write e_1 for the edge of D that originates in v_1, $A_1 = A_{e_1}$, and $v_2 = A_1(v_1)$. Note that $e_1' = A_1(e_1)$ is the edge that terminates in v_2. We do the same with v_2, defining $v_3 = A_2(v_2)$, and continue in this way until we arrive at $v_1 = A_n(v_n)$ and e_n' the edge of D terminating in v_1 This will occur after finitely many steps (even though G need not be finitely generated), so the cycle $\{v_1, v_2, \ldots, v_n\}$ containing v_1 always exists.

The element

$$E_1 = A_n \circ A_{n-1} \circ \ldots \circ A_2 \circ A_1$$

generates the stabilizer of v_1 in G. It is elliptic or the identity, and its order $m \geqq 1$ satisfies

$$m \sum_{k=1}^{n} \alpha_k = 2\pi$$

where α_k is the interior angle of D at v_k. Recall that this is because the translate domains $A_n(D), A_nA_{n-1}(D), \ldots, E_1(D)$ share the vertex v_1 and cover the sector of the neighborhood of v_1 of angle $2\pi/m$ lying between e_n' and $E_1(e_n')$.

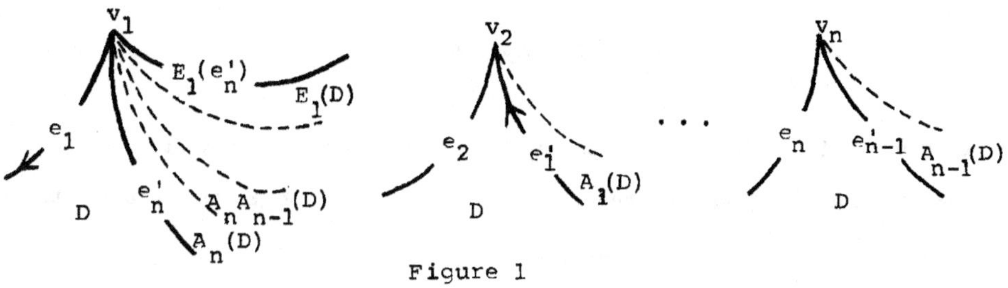

Figure 1

In the sequel we will use the letters v_k, α_k, m, n, etc., generically although their values may vary from cycle to cycle.

1.2. If f is a topological deformation of G which induces the homomorphism χ, we will write $\hat{v}_k = f(v_k)$, $\hat{e}_k = f(e_k)$, $\hat{A}_k = \chi(A_k)$, etc. (Note that $\hat{e}_k$ may not have a well-defined tangent at its endpoints, so $\hat{\alpha}_k$ may not be defined.) The process described above for defining a cycle can be carried out just as well in the image domain beginning with $\hat{v}_1$, and will lead to $\{\hat{v}_1, \hat{v}_2, \ldots, \hat{v}_n\}$ since

$$\hat{A}_k(\hat{e}_k) = f(e_k') = \hat{e}_k'$$

as a consequence of (1). Thus the neighborhood of $\hat{v}_1$ is covered by the images under f of the corners of D at $v_1, \ldots, v_n$ translated by the corresponding elements of

$\chi(G)$. In particular, when f is restricted to a neighborhood of v_1 in which it is an orientation-preserving homeomorphism, the parts of the m circular arcs $e_n', E_1(e_n'), \ldots, E_1^{m-1}(e_n')$ are sent respectively to parts of the curves $\hat{e}_n', \hat{E}_1(\hat{e}_n'), \ldots, \hat{E}_1^{m-1}(\hat{e}_n')$ which come together at $\hat{v}_1$ and occur in the same cyclic order as their counterparts in Δ.

2. It is easy to see the necessity of the condition stated in the Theorem. Since $\hat{E}_1$ effects a cyclic permutation of the curves listed in the preceding paragraph, it is locally a rotation of angle $2\pi/m$ about v_1 and so

$$\operatorname{tr} \hat{E}_1 = 2 \cos(\pi/m) = \operatorname{tr} E_1 .$$

Every elliptic $E \in G$ is conjugate to an element of G fixing some vertex v_1 of D, and that element is a power of the corresponding E_1. Since $\hat{E}$ is conjugate to the same power of $\hat{E}_1$, we conclude that $\operatorname{tr} \hat{E} = \operatorname{tr} E$.

3. We will now prove the sufficiency of the condition by constructing $f : D \to \hat{\mathbb{C}}$ with the following properties:

(i) f is a local homeomorphism in D;
(ii) for each edge e of D, $\hat{e} = f(e)$ is a smooth (C^1) curve and $f(A_e(e)) = \hat{A}_e(\hat{e})$;
(iii) for each cycle of vertices of D, $\sum_{k=1}^{n} \hat{\alpha}_k = 2\pi/m$.

When we say $\hat{e}$ is smooth we include of course any vertices it may possess. By convention we will always assume that $0 \leq \hat{\alpha}_k < 2\pi$.

To see that these properties are indeed sufficient, we note that (ii) permits us to extend f to all of Δ via the group action; in other words, for each $z \in \Delta$ fin[d] $A \in G$ so that $A(z) \in D$ and define

$$f(z) = \hat{A}^{-1} f A(z)$$

which does not depend upon the choice of A. We then see that f is a local homeomorphism at z by applying (i), (ii), or (iii) in the cases that $A(z)$ is an interior point of D, an interior point of an edge of D, or a vertex of D respectively.

3.1. Naturally the construction begins at the vertices. Given a cycle $\{v_1,\ldots,v_n\}$, when $m = 1$ we may choose $\hat{v}_1 \in \hat{\mathbb{C}}$ arbitrarily. When $m = 2$ we must take $\hat{v}_1$ to be either fixed point of the elliptic element $\hat{E}_1$. When $m > 2$ we must take $\hat{v}_1$ to be the unique fixed point of $\hat{E}_1$ satisfying the following property: there is a $B \in \mathrm{Aut}\,\hat{\mathbb{C}}$ such that $B(\hat{v}_1) = 0$ and $B\hat{E}B^{-1}(z) = e^{2\pi i/m}z$. We extend f to the rest of the cycle via the group action: $\hat{v}_{k+1} = \hat{A}_k(\hat{v}_k)$. This is to be done for each cycle. Since the vertices of D form a countable discrete subset of Δ, the resulting f is continuous.

3.2. Now let e be an edge of D, beginning and ending in the points v, v' respectively. We provisionally choose as $\hat{e}$ any immersed smooth curve in $\hat{\mathbb{C}}$, with the condition that if v (or v') is a vertex then $\hat{e}$ begins (or ends) at $\hat{v}$ (or $\hat{v}'$). It is important that the direction of $\hat{e}$ exist at a vertex; however, we do not know at this moment what we want that direction to be, and at a later step $\hat{e}$ may be modified.

Define f along e so that it is smooth and its image is $\hat{e}$. Define f along e' so that it satisfies (ii). Do this for each pair e, e' of edges of D and give each image edge $\hat{e}$ the orientation carried over by f from e. Since finitely many edges meet any compact subset of Δ, f is continuous on the union of the edges of D.

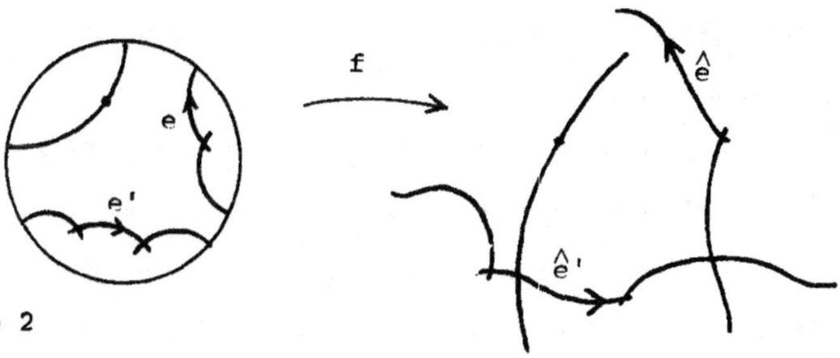

Figure 2

We have done nothing to ensure the angle condition (iii). Nonetheless, for each cycle we have automatically

(iv) $\sum_{k=1}^{n} \hat{\alpha}_k \equiv 2\pi/m$ (modulo 2π).

Indeed, the $n+1$ image edges $\hat{e}_n' = \hat{A}_n(\hat{e}_n), \hat{A}_n\hat{A}_{n-1}(\hat{e}_{n-1}), \hat{A}_n\hat{A}_{n-1}\hat{A}_{n-2}(\hat{e}_{n-2}), \ldots,$ $\hat{E}_1(\hat{e}_n), \hat{E}_1(\hat{e}_n')$ meet at $\hat{v}_1$ with the n angles $\hat{\alpha}_n, \hat{\alpha}_{n-1}, \hat{\alpha}_{n-2}, \ldots, \hat{\alpha}_1$ successively. By the conditions imposed on $\hat{E}_1$ and $\hat{v}_1$, the angle between $\hat{e}_n'$ and $\hat{E}_1(\hat{e}_n')$ is $2\pi/m$. Hence (iv) holds.

We will modify the image edges by means of the following procedure. Let $\hat{e}$ originate in $\hat{v}$. Rotate an initial portion of $\hat{e}$ counterclockwise about $\hat{v}$ by an angle of β, then join this rotated piece smoothly to the remainder of $\hat{e}$. The new curve will again be called $\hat{e}$, we redefine $\hat{e}' = \hat{A}_e(e)$, and adjust the piecewise smooth immersion f on e, e' accordingly. This process will be called "rotation of $\hat{e}$ by β near $\hat{v}$."

Figure 3

Note that since $\hat{A}_e$ is a conformal mapping, a rotation of $\hat{e}$ by β near its initial point involves a rotation of $\hat{e}'$ also by β near its terminal point. In this way the condition (iv) is preserved.

By means of rotations it is easy to achieve (iii) for any given cycle. First suppose $\hat{\alpha}_k = 0$. If also $\hat{\alpha}_{k+1} = 0$ then a rotation of $\hat{e}_k$ near $\hat{v}_k$ by any positive angle will suffice to make both $\hat{\alpha}_k$ and $\hat{\alpha}_{k+1}$ nonzero without affecting any other angles. If on the other hand $\hat{\alpha}_{k+1} \neq 0$ we accomplish the same thing rotating $\hat{e}_k$ b

$\hat{\alpha}_{k+1}/2$ near $\hat{v}_k$. Thus we make $\hat{\alpha}_k > 0$ for $k = 1,2,\ldots,n$.

Now to make f satisfy (iii) write

$$\sum_{k=1}^{n} \hat{\alpha}_k = 2\pi(p+1/m)$$

for some integer $p \geq 0$. The case $n = 1$ is rather trivial and will be put aside. Let $\delta = 2\pi/mn$. The first step is as follows: if $\hat{\alpha}_1 \leq \delta$, do nothing; otherwise rotate $\hat{e}_1$ by $\hat{\alpha}_1 - \delta$ near $\hat{v}_1$. The succeeding steps are the same: if $\hat{\alpha}_k \leq \delta$ (note $\hat{\alpha}_k$ has just been redefined) do nothing; otherwise rotate $\hat{e}_k$ by $\hat{\alpha}_k - \delta$ near $\hat{v}_k$. After $n-1$ steps we have $\hat{\alpha}_k \leq \delta$ for $k = 1,2,\ldots,n-1$ and so $\Sigma\hat{\alpha}_k \leq (n-1)\delta + 2\pi < 2\pi(1+1/m)$. This implies $p = 0$, which is what (iii) asserts.

We must do the above for every cycle. Since each edge will be altered at most a finite number of times, this process will assure a well defined f on the union of the edges of D.

3.3. It remains to define f in the interior of D. We will lose no generality by supposing that D is the Ford fundamental domain for G and thus is starshaped with respect to 0. Since Δ/G is by hypothesis not compact, D is an unbounded set in the hyperbolic metric; i.e., the euclidean closure of D does not lie in Δ. In particular, every annulus $r \leq |z| \leq r'$ in Δ meets D.

Choose a sequence $\{r_j\}$ strictly increasing to 1, and consider the sets

$$D_1 = \{z \in D : |z| \leq r_1\},$$

$$D_j = \{z \in D : r_{j-1} \leq |z| \leq r_j\},$$

for $j \geq 2$. We require r_1 large enough so that D_1 includes part of the boundary of D.

Now D_1 is a Jordan domain whose boundary is a fintie union of arcs of the circle $|z| = r_1$ together with subarcs of edges of D. Each bounded connected component of $D-D_1$ is a similar type of Jordan domain, abutting D_1 along a single arc of $|z| = r_1$. Therefore the set

$$\mathcal{U}_1 = D_1 \cup \bigcup \text{bounded components of } D-D_1$$

is again a Jordan domain. We may decompose the boundary of $\mathcal{U}_1$ as

$$\partial\mathcal{U}_1 = \sigma \cup \tau$$

where τ is an arc of $|z| = r_1$ (this arc τ exists because $D-D_1$ has at least one unbounded component). We have already defined f on the part of σ formed from edges of D, and we extend f to the rest of σ subject to the condition that $\hat{\sigma} = f(\sigma)$ be an immersed piecewise smooth oriented one-dimensional submanifold (with boundary) of $\hat{\mathbb{C}}$. Thus $\hat{\sigma}$ has an immersed tubular neighborhood and we are able to extend f to a local hemeomorphism from $\mathcal{U}_1$ onto the appropriately chosen "half heighborhood" determined by the orientation. (Note that τ will be sent to a curve $\hat{\tau}$ running back to $\hat{\sigma}$.)

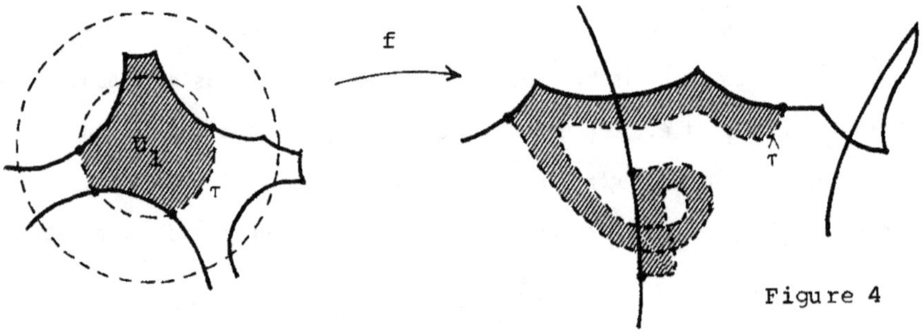

Figure 4

The next step is to define

$$\mathcal{U}_2 = (D_2-\mathcal{U}_1) \cup \bigcup \text{bounded components of } D-(D_2 \cup \mathcal{U}_1).$$

The same reasoning as that just used can be applied to each *component* of $\mathcal{U}_2$. These are Jordan domains each of whose boundaries contains an arc of $|z| = r_2$. Sinc f has not yet been defined on these arcs, we can extend it to a local homeomorphism in $\mathcal{U}_1 \cup \mathcal{U}_2$. Continuing in this manner we set

$$\mathcal{U}_j = \left(D_j - \bigcup_{i=1}^{j-1} \mathcal{U}_i\right) \cup \bigcup \text{bounded components of } D - \left(D_j \cup \bigcup_{i=1}^{j-1} \mathcal{U}_i\right)$$

and obtain the desired local homeomorphism in $D = \bigcup \mathcal{U}_j$. This completes the proof of the theorem.

4. The Theorem is also true when modified to require the topological deformation to reverse orientation. The only change in the proof is in 3.1 where we take $B(\hat{v}_1) = \infty$ instead of 0.

5. A stronger result should be possible for finitely generated groups. For these groups the algebraic variety of homomorphisms from G into $\mathrm{Aut}\,\hat{\mathbb{C}}$ is finite dimensional while the space of projective structures on the noncompact Riemann surface Δ/G (canonically in one-to-one correpsondence with the vector space of holomorphic quadratic differentials on Δ/G) is infinite dimensional. Thus we conjecture that for these groups the Theorem holds when "topological deformation" is replaced by "holomorphic deformation."

I would like to thank D. Gallo and A. Verjovsky for many helpful comments.

CENTRO DE INVESTIGACION Y DE
ESTUDIOS AVANZADOS DEL IPN
MEXICO, D.F., MEXICO

REFERENCES

[1] L. Ford, *Automorphic Functions*, Chelsea Publishing Company, 1929, 1957 reprinting.

[2] R.C. Gunning, "Special coordinate covers on Riemann surfaces," *Math. Ann.* 190 (1967), 67-86.

[3] R.C. Gunning, "Affine and projective structures on Riemann surfaces," Proceedings of the 1978 Stony Brook Conference, *Ann. of Math. Studies* 97 (1981), 225-244.

[4] D.A. Hejhal, "Monodromy groups and Poincaré series," *Bull.* AMS 84 (1978), 339-376.

Vol. 817: L. Gerritzen, M. van der Put, Schottky Groups and Mumford Curves. VIII, 317 pages. 1980.

Vol. 818: S. Montgomery, Fixed Rings of Finite Automorphism Groups of Associative Rings. VII, 126 pages. 1980.

Vol. 819: Global Theory of Dynamical Systems. Proceedings, 1979. Edited by Z. Nitecki and C. Robinson. IX, 499 pages. 1980.

Vol. 820: W. Abikoff, The Real Analytic Theory of Teichmüller Space. VII, 144 pages. 1980.

Vol. 821: Statistique non Paramétrique Asymptotique. Proceedings, 1979. Edited by J.-P. Raoult. VII, 175 pages. 1980.

Vol. 822: Séminaire Pierre Lelong-Henri Skoda, (Analyse) Années 1978/79. Proceedings. Edited by P. Lelong et H. Skoda. VIII, 356 pages. 1980.

Vol. 823: J. Král, Integral Operators in Potential Theory. III, 171 pages. 1980.

Vol. 824: D. Frank Hsu, Cyclic Neofields and Combinatorial Designs. VI, 230 pages. 1980.

Vol. 825: Ring Theory, Antwerp 1980. Proceedings. Edited by F. van Oystaeyen. VII, 209 pages. 1980.

Vol. 826: Ph. G. Ciarlet et P. Rabier, Les Equations de von Kármán. VI, 181 pages. 1980.

Vol. 827: Ordinary and Partial Differential Equations. Proceedings, 1978. Edited by W. N. Everitt. XVI, 271 pages. 1980.

Vol. 828: Probability Theory on Vector Spaces II. Proceedings, 1979. Edited by A. Weron. XIII, 324 pages. 1980.

Vol. 829: Combinatorial Mathematics VII. Proceedings, 1979. Edited by R. W. Robinson et al.. X, 256 pages. 1980.

Vol. 830: J. A. Green, Polynomial Representations of GL_n. VI, 118 pages. 1980.

Vol. 831: Representation Theory I. Proceedings, 1979. Edited by V. Dlab and P. Gabriel. XIV, 373 pages. 1980.

Vol. 832: Representation Theory II. Proceedings, 1979. Edited by V. Dlab and P. Gabriel. XIV, 673 pages. 1980.

Vol. 833: Th. Jeulin, Semi-Martingales et Grossissement d'une Filtration. IX, 142 Seiten. 1980.

Vol. 834: Model Theory of Algebra and Arithmetic. Proceedings, 979. Edited by L. Pacholski, J. Wierzejewski, and A. J. Wilkie. VI, 10 pages. 1980.

ol. 835: H. Zieschang, E. Vogt and H.-D. Coldewey, Surfaces and lanar Discontinuous Groups. X, 334 pages. 1980.

ol. 836: Differential Geometrical Methods in Mathematical Physics. roceedings, 1979. Edited by P. L. García, A. Pérez-Rendón, and M. Souriau. XII, 538 pages. 1980.

ol. 837: J. Meixner, F. W. Schäfke and G. Wolf, Mathieu Functions nd Spheroidal Functions and their Mathematical Foundations urther Studies. VII, 126 pages. 1980.

ol. 838: Global Differential Geometry and Global Analysis. Pro-edings 1979. Edited by D. Ferus et al. XI, 299 pages. 1981.

ol. 839: Cabal Seminar 77 – 79. Proceedings. Edited by A. S. chris, D. A. Martin and Y. N. Moschovakis. V, 274 pages. 1981.

l. 840: D. Henry, Geometric Theory of Semilinear Parabolic Equa-ns. IV, 348 pages. 1981.

. 841: A. Haraux, Nonlinear Evolution Equations- Global Behaviour Solutions. XII, 313 pages. 1981.

. 842: Séminaire Bourbaki vol. 1979/80. Exposés 543–560. 317 pages. 1981.

. 843: Functional Analysis, Holomorphy, and Approximation ory. Proceedings. Edited by S. Machado. VI, 636 pages. 1981.

844: Groupe de Brauer. Proceedings. Edited by M. Kervaire and Ojanguren. VII, 274 pages. 1981.

Vol. 845: A. Tannenbaum, Invariance and System Theory: Algebraic and Geometric Aspects. X, 161 pages. 1981.

Vol. 846: Ordinary and Partial Differential Equations, Proceedings. Edited by W. N. Everitt and B. D. Sleeman. XIV, 384 pages. 1981.

Vol. 847: U. Koschorke, Vector Fields and Other Vector Bundle Morphisms – A Singularity Approach. IV, 304 pages. 1981.

Vol. 848: Algebra, Carbondale 1980. Proceedings. Ed. by R. K. Amayo. VI, 298 pages. 1981.

Vol. 849: P. Major, Multiple Wiener-Itô Integrals. VII, 127 pages. 1981.

Vol. 850: Séminaire de Probabilités XV. 1979/80. Avec table générale des exposés de 1966/67 à 1978/79. Edited by J. Azéma and M. Yor. IV, 704 pages. 1981.

Vol. 851: Stochastic Integrals. Proceedings, 1980. Edited by D. Williams. IX, 540 pages. 1981.

Vol. 852: L. Schwartz, Geometry and Probability in Banach Spaces. X, 101 pages. 1981.

Vol. 853: N. Boboc, G. Bucur, A. Cornea, Order and Convexity in Potential Theory: H-Cones. IV, 286 pages. 1981.

Vol. 854: Algebraic K-Theory. Evanston 1980. Proceedings. Edited by E. M. Friedlander and M. R. Stein. V, 517 pages. 1981.

Vol. 855: Semigroups. Proceedings 1978. Edited by H. Jürgensen, M. Petrich and H. J. Weinert. V, 221 pages. 1981.

Vol. 856: R. Lascar, Propagation des Singularités des Solutions d'Equations Pseudo-Différentielles à Caractéristiques de Multiplicités Variables. VIII, 237 pages. 1981.

Vol. 857: M. Miyanishi. Non-complete Algebraic Surfaces. XVIII, 244 pages. 1981.

Vol. 858: E. A. Coddington, H. S. V. de Snoo: Regular Boundary Value Problems Associated with Pairs of Ordinary Differential Expressions. V, 225 pages. 1981.

Vol. 859: Logic Year 1979–80. Proceedings. Edited by M. Lerman, J. Schmerl and R. Soare. VIII, 326 pages. 1981.

Vol. 860: Probability in Banach Spaces III. Proceedings, 1980. Edited by A. Beck. VI, 329 pages. 1981.

Vol. 861: Analytical Methods in Probability Theory. Proceedings 1980. Edited by D. Dugué, E. Lukacs, V. K. Rohatgi. X, 183 pages. 1981.

Vol. 862: Algebraic Geometry. Proceedings 1980. Edited by A. Libgober and P. Wagreich. V, 281 pages. 1981.

Vol. 863: Processus Aléatoires à Deux Indices. Proceedings, 1980. Edited by H. Korezlioglu, G. Mazziotto and J. Szpirglas. V, 274 pages. 1981.

Vol. 864: Complex Analysis and Spectral Theory. Proceedings, 1979/80. Edited by V. P. Havin and N. K. Nikol'skii. VI, 480 pages. 1981.

Vol. 865: R. W. Bruggeman, Fourier Coefficients of Automorphic Forms. III, 201 pages. 1981.

Vol. 866: J.-M. Bismut, Mécanique Aléatoire. XVI, 563 pages. 1981.

Vol. 867: Séminaire d'Algèbre Paul Dubreil et Marie-Paule Malliavin. Proceedings, 1980. Edited by M.-P. Malliavin. V, 476 pages. 1981.

Vol. 868: Surfaces Algébriques. Proceedings 1976–78. Edited by J. Giraud, L. Illusie et M. Raynaud. V, 314 pages. 1981.

Vol. 869: A. V. Zelevinsky, Representations of Finite Classical Groups. IV, 184 pages. 1981.

Vol. 870: Shape Theory and Geometric Topology. Proceedings, 1981. Edited by S. Mardešić and J. Segal. V, 265 pages. 1981.

Vol. 871: Continuous Lattices. Proceedings, 1979. Edited by B. Banaschewski and R.-E. Hoffmann. X, 413 pages. 1981.

Vol. 872: Set Theory and Model Theory. Proceedings, 1979. Edited by R. B. Jensen and A. Prestel. V, 174 pages. 1981.

Vol. 873: Constructive Mathematics, Proceedings, 1980. Edited by F. Richman. VII, 347 pages. 1981.

Vol. 874: Abelian Group Theory. Proceedings, 1981. Edited by R. Göbel and E. Walker. XXI, 447 pages. 1981.

Vol. 875: H. Zieschang, Finite Groups of Mapping Classes of Surfaces. VIII, 340 pages. 1981.

Vol. 876: J. P. Bickel, N. El Karoui and M. Yor. Ecole d'Eté de Probabilités de Saint-Flour IX – 1979. Edited by P. L. Hennequin. XI, 280 pages. 1981.

Vol. 877: J. Erven, B.-J. Falkowski, Low Order Cohomology and Applications. VI, 126 pages. 1981.

Vol. 878: Numerical Solution of Nonlinear Equations. Proceedings, 1980. Edited by E. L. Allgower, K. Glashoff, and H.-O. Peitgen. XIV, 440 pages. 1981.

Vol. 879: V. V. Sazonov, Normal Approximation – Some Recent Advances. VII, 105 pages. 1981.

Vol. 880: Non Commutative Harmonic Analysis and Lie Groups. Proceedings, 1980. Edited by J. Carmona and M. Vergne. IV, 553 pages. 1981.

Vol. 881: R. Lutz, M. Goze, Nonstandard Analysis. XIV, 261 pages. 1981.

Vol. 882: Integral Representations and Applications. Proceedings, 1980. Edited by K. Roggenkamp. XII, 479 pages. 1981.

Vol. 883: Cylindric Set Algebras. By L. Henkin, J. D. Monk, A. Tarski, H. Andréka, and I. Németi. VII, 323 pages. 1981.

Vol. 884: Combinatorial Mathematics VIII. Proceedings, 1980. Edited by K. L. McAvaney. XIII, 359 pages. 1981.

Vol. 885: Combinatorics and Graph Theory. Edited by S. B. Rao. Proceedings, 1980. VII, 500 pages. 1981.

Vol. 886: Fixed Point Theory. Proceedings, 1980. Edited by E. Fadell and G. Fournier. XII, 511 pages. 1981.

Vol. 887: F. van Oystaeyen, A. Verschoren, Non-commutative Algebraic Geometry, VI, 404 pages. 1981.

Vol. 888: Padé Approximation and its Applications. Proceedings, 1980. Edited by M. G. de Bruin and H. van Rossum. VI, 383 pages. 1981.

Vol. 889: J. Bourgain, New Classes of $\mathcal{L}^p$-Spaces. V, 143 pages. 1981.

Vol. 890: Model Theory and Arithmetic. Proceedings, 1979/80. Edited by C. Berline, K. McAloon, and J.-P. Ressayre. VI, 306 pages. 1981.

Vol. 891: Logic Symposia, Hakone, 1979, 1980. Proceedings, 1979, 1980. Edited by G. H. Müller, G. Takeuti, and T. Tugué. XI, 394 pages. 1981.

Vol. 892: H. Cajar, Billingsley Dimension in Probability Spaces. III, 106 pages. 1981.

Vol. 893: Geometries and Groups. Proceedings. Edited by M. Aigner and D. Jungnickel. X, 250 pages. 1981.

Vol. 894: Geometry Symposium. Utrecht 1980, Proceedings. Edited by E. Looijenga, D. Siersma, and F. Takens. V, 153 pages. 1981.

Vol. 895: J.A. Hillman, Alexander Ideals of Links. V, 178 pages. 1981.

Vol. 896: B. Angéniol, Familles de Cycles Algébriques – Schéma de Chow. VI, 140 pages. 1981.

Vol. 897: W. Buchholz, S. Feferman, W. Pohlers, W. Sieg, Iterated Inductive Definitions and Subsystems of Analysis: Recent Proof-Theoretical Studies. V, 383 pages. 1981.

Vol. 898: Dynamical Systems and Turbulence, Warwick, 1980. Proceedings. Edited by D. Rand and L.-S. Young. VI, 390 pages. 1981.

Vol. 899: Analytic Number Theory. Proceedings, 1980. Edited by M.I. Knopp. X, 478 pages. 1981.

Vol. 900: P. Deligne, J. S. Milne, A. Ogus, and K.-Y. Shih, Hodge Cycles, Motives, and Shimura Varieties. V, 414 pages. 1982.

Vol. 901: Séminaire Bourbaki vol. 1980/81 Exposés 561–578. III, 299 pages. 1981.

Vol. 902: F. Dumortier, P.R. Rodrigues, and R. Roussarie, Germs of Diffeomorphisms in the Plane. IV, 197 pages. 1981.

Vol. 903: Representations of Algebras. Proceedings, 1980. Edited by M. Auslander and E. Lluis. XV, 371 pages. 1981.

Vol. 904: K. Donner, Extension of Positive Operators and Korovkin Theorems. XII, 182 pages. 1982.

Vol. 905: Differential Geometric Methods in Mathematical Physics. Proceedings, 1980. Edited by H.-D. Doebner, S.J. Andersson, and H.R. Petry. VI, 309 pages. 1982.

Vol. 906: Séminaire de Théorie du Potentiel, Paris, No. 6. Proceedings. Edité par F. Hirsch et G. Mokobodzki. IV, 328 pages. 1982.

Vol. 907: P. Schenzel, Dualisierende Komplexe in der lokalen Algebra und Buchsbaum-Ringe. VII, 161 Seiten. 1982.

Vol. 908: Harmonic Analysis. Proceedings, 1981. Edited by F. Ricci and G. Weiss. V, 325 pages. 1982.

Vol. 909: Numerical Analysis. Proceedings, 1981. Edited by J.P. Hennart. VII, 247 pages. 1982.

Vol. 910: S.S. Abhyankar, Weighted Expansions for Canonical Desingularization. VII, 236 pages. 1982.

Vol. 911: O.G. Jørsboe, L. Mejlbro, The Carleson-Hunt Theorem on Fourier Series. IV, 123 pages. 1982.

Vol. 912: Numerical Analysis. Proceedings, 1981. Edited by G. A. Watson. XIII, 245 pages. 1982.

Vol. 913: O. Tammi, Extremum Problems for Bounded Univalent Functions II. VI, 168 pages. 1982.

Vol. 914: M. L. Warshauer, The Witt Group of Degree k Maps and Asymmetric Inner Product Spaces. IV, 269 pages. 1982.

Vol. 915: Categorical Aspects of Topology and Analysis. Proceedings, 1981. Edited by B. Banaschewski. XI, 385 pages. 1982.

Vol. 916: K.-U. Grusa, Zweidimensionale, interpolierende Lg-Splines und ihre Anwendungen. VIII, 238 Seiten. 1982.

Vol. 917: Brauer Groups in Ring Theory and Algebraic Geometry. Proceedings, 1981. Edited by F. van Oystaeyen and A. Verschoren. VIII, 300 pages. 1982.

Vol. 918: Z. Semadeni, Schauder Bases in Banach Spaces of Continuous Functions. V, 136 pages. 1982.

Vol. 919: Séminaire Pierre Lelong – Henri Skoda (Analyse) Années 1980/81 et Colloque de Wimereux, Mai 1981. Proceedings. Edité par P. Lelong et H. Skoda. VII, 383 pages. 1982.

Vol. 920: Séminaire de Probabilités XVI, 1980/81. Proceedings Edité par J. Azéma et M. Yor. V, 622 pages. 1982.

Vol. 921: Séminaire de Probabilités XVI, 1980/81. Supplément Géométrie Différentielle Stochastique. Proceedings. Edité par J. Azéma et M. Yor. III, 285 pages. 1982.

Vol. 922: B. Dacorogna, Weak Continuity and Weak Lower Semicontinuity of Non-Linear Functionals. V, 120 pages. 1982.

Vol. 923: Functional Analysis in Markov Processes. Proceedings 1981. Edited by M. Fukushima. V, 307 pages. 1982.

Vol. 924: Séminaire d'Algèbre Paul Dubreil et Marie-Paule Malliavin Proceedings, 1981. Edité par M.-P. Malliavin. V, 461 pages. 198

Vol. 925: The Riemann Problem, Complete Integrability and Arithmetic Applications. Proceedings, 1979-1980. Edited by D. Chudnovsky and G. Chudnovsky. VI, 373 pages. 1982.

Vol. 926: Geometric Techniques in Gauge Theories. Proceedings 1981. Edited by R. Martini and E.M.de Jager. IX, 219 pages. 198